09273

DESIGN PROCEDURES FOR HYDROSTATIC BEARINGS

Design Procedures for Hydrostatic Bearings

by **W. B. ROWE**

B.Sc., Ph.D., C.Eng., M.I.Mech.E., M.I.Prod.E.
Principal Lecturer in Production Engineering,
Lanchester Polytechnic, Coventry

and **J. P. O'DONOGHUE**

B.E., Ph.D., C.Eng., M.I.Mech.E., Assoc. Member A.S.M.E.
Principal Lecturer in Mechanical Engineering,
Lanchester Polytechnic, Coventry

THE MACHINERY PUBLISHING CO. LTD.

Head Office: NEW ENGLAND HOUSE, NEW ENGLAND STREET,
BRIGHTON, SUSSEX, BN1 4HN

Reg. Office: CLIFTON HOUSE, 83-117 EUSTON ROAD, LONDON, N.W.1

First Published 1971

Authors: W. B. Rowe & J. P. O'Donoghue

ISBN 0 85333 224 X

Printed in Great Britain by
Graphic Art Services (Brighton) Ltd.

Preface

This book contains eight design procedures for a range of hydrostatic bearing applications. The authors make no apology for not discussing underlying theory in detail since numerous papers are available in specialised journals for those who wish to enquire further. The obvious and outstanding need expressed by many designers was for step by step guidance towards safe design practice.

The book developed from an article published in *Machinery and Production Engineering*. The first article contained the procedure for hydrostatic journal bearings and appeared to be well received. Furthermore, as subsequent design procedures were published in a similar manner we were encouraged by receiving communications from various organisations, many of whom had successfully designed bearings without previous specialised knowledge. They had managed to follow the instructions and thereby achieve a satisfactory solution. Inevitably there were some whose particular applications involved critical problems such as a very wide operating range of temperatures. Some design problems cannot be solved by simple procedures and it is therefore suggested that specialists should be consulted.

The design procedures owe much to ideas drawn from various sources, of which the principal papers have been referenced. Particular acknowledgements are due to Mr R L Aston, Head of the Production Engineering Department at Lanchester Polytechnic, who was involved in the original work on conical hydrostatic bearings and was co-author of the corresponding article; also to Mr R S Wearing, Lecturer in the Mechanical Engineering Department, Lanchester Polytechnic, who made a study of bearings involving the Yates principle and was co-author of the combined journal-thrust bearing article.

W B Rowe
J P O'Donoghue

Contents

Chapter One

Hydrostatic journal bearings

W B ROWE and J P O'DONOGHUE

An essential feature of an externally pressurized bearing is a compensating mechanism, the purpose of which is to allow the pressure in the recesses to be adjusted in response to an external load. A common type of hydrostatic bearing is represented in Fig. 1. In this bearing, four recesses are supplied with oil, each through a separate capillary tube. In other types of bearing, compensation is provided alternatively by orifices, by a separate constant flow valve or pump for each recess, or by diaphragm valves.

The designer specifies the dimensions for the bearing pads first and then carries out a procedure for the restrictor. Often the designer will choose capillary compensation because the journal position is then independent of temperature and manufacture is simple. Orifice compensation however yields a stiffer bearing than does capillary compensation, and an added advantage is that the flow rate is less affected by changes in temperature. With the constant flow compensation type of bearing, stiffness is further improved, and changes in flow rate are further reduced. With "diaphragm compensation", it is possible to achieve infinite bearing stiffness by suitable choice of the diaphragm stiffness.

The flow rates predicted by the expressions given in the procedure are accurate for thin land bearings but result in slight over-estimates for thick land bearings. However, the designer will usually provide for at least a 20 per cent excess of pump capacity over the predicted value, to allow for manufacturing tolerances and variations in temperature.

In the following procedure, the designer specifies the extreme load that the bearing must support and the speed at which the bearing will operate. Values of stiffness and flow rate are calculated and guidance is given in the choice of bearing dimensions. However, the most important feature of the procedure is that by following the steps outlined the minimum power condition is always obtained. A computer-aided solution, using an exact theory, is also available and further information on this aspect may be obtained from the authors.

If the manufacturing tolerances are wide, and large bearing clearances are accordingly selected, the total power expended in the bearing will be larger than if smaller clearances can be employed. In either instance, however, it is still

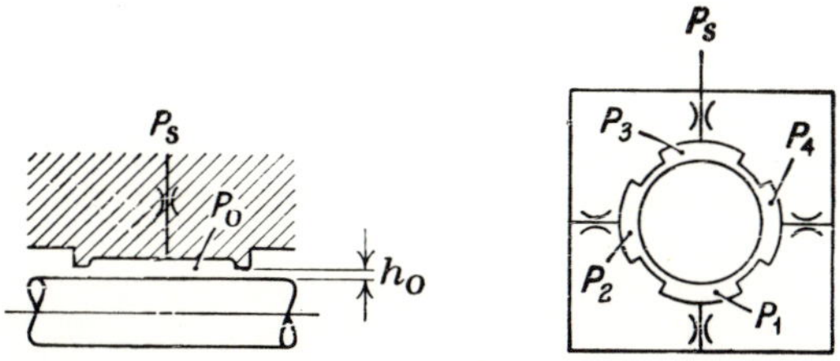

Fig 1

Diagrammatic representation of a typical 4-pad hydrostatic journal bearing

Fig 2

Representations of thin-land hydrostatic journal bearings with (x) and without (y) axial exhaust slots. The latter design is that recommended

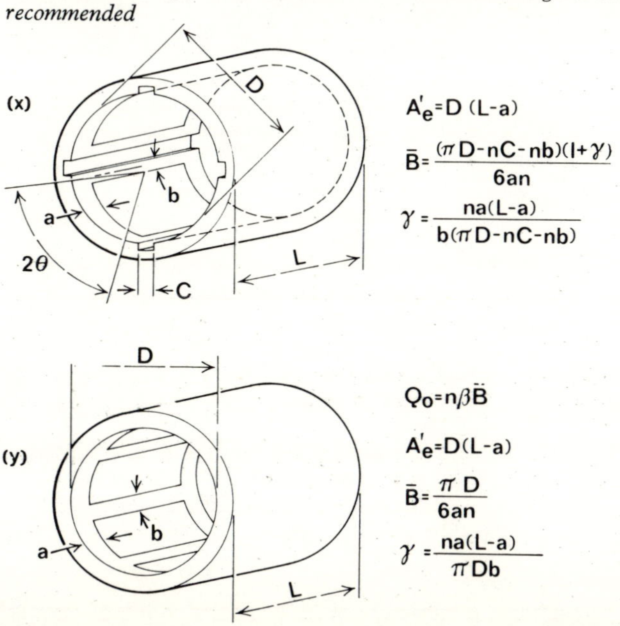

important that the minimum power condition should be achieved, in order to avoid high temperatures and excessive heat dissipation.

Two types of bearing construction are shown in Fig. 2. In the design seen at (x), the bearing recesses are divided by axially-extending exhaust slots. When a load is applied to this bearing, the resulting deflection is directly in line with the applied load, which is not true of the bearing shown at (y), at high speeds and eccentricity ratios. Also it is possible to achieve a higher flow rate through the bearing (x) for a given bearing gap. For most applications, however, the bearing (y), without axial slots, is recommended, because the friction area is lower and manufacture is simpler. In addition, slots may cause air to be drawn into the bearing.

The power minimization procedure is based on making the pumping power equal to the power dissipated due to the sliding speed of the bearing. It can be shown theoretically that this condition is always within 15 per cent of the minimum required, and is the actual minimum if the procedure is followed exactly.

Basis of the design procedure

The derivation of the equations for journal bearing stiffness has been given in a previous paper[(2)] [all references at end of chapter] and a discussion of the minimum power condition has been published by Opitz [(3)]. The fundamental principles underlying the design of externally pressurized bearings are discussed in much greater detail in an article by Rowe and O'Donoghue [(4)]. Other relevant papers are listed [(1)] [(5)].

Bearing pad design procedure

(1) Specify extreme load W'.

(2) Specify number of bearing recesses n.

For high precision bearings which do not exhibit directional stiffness properties, the value of n should be 6. In most cases, n will be 4 and manufacturing will be simpler.

(3) Specify the design parameter $\beta = Pro/p_s$

Choosing $\beta = 0{\cdot}5$ yields an efficient bearing with high stiffness. β should not be greater than 0·5, but may be reduced on a given bearing which has already been made to 0·2 in order to reduce the flow rate.

(4) Specify journal diameter D.

A convenient guide is D (minimum) $= \sqrt{0{\cdot}015W'}$ when working in units of lbf. and inches, since this leads to a supply pressure which should not be greatly in excess of 300 lbf. per sq. in.

(5) Specify bearing length L.

Make $L = D$ unless there are limitations of space.

(6) Specify the bearing clearance h_o.

For minimum power and high stiffness, the clearance should be as small as possible consistent with manufacturing accuracy, that is, within the

Symbols

P_s Supply pressure, or maximum obtainable pressure
P_r Recess pressure
o Suffix denoting design condition
h Bearing gap
D Bearing diameter
L Bearing length
a Width of axial flow land
b Width of circumferential flow land
c Width of axial exhaust slot
W' Extreme bearing load
2θ Angle subtended by one pad and recess
β Design parameter $= Pro/P_s$
γ $\dfrac{\text{Axial flow resistance}}{\text{Circumferential flow resistance}}$ = circumferential flow factor
λ Stiffness
$\bar{\lambda}$ Dimensionless stiffness $= \dfrac{\lambda h_o}{P_s D(L - a)}$
$\bar{\lambda}_o{}^*$ Design value from capillary equation with $\beta = 0{\cdot}5$
A_L Land area surrounding one recess only
A_R Recess area for one recess only
A_f Effective friction area $= \dfrac{A_R}{4} + A_L$
$\bar{B}$ Flow shape factor
v Bearing sliding speed
ΔT Maximum temperature rise when oil flows through bearing once
ρ Mass density of oil
$S.G.$ Specific gravity of oil
K_c Capillary coefficient
K_{or} Orifice constant
C_f Discharge coefficient
N_t' Transition speed from laminar to turbulent conditions
ω Raimondi and Boyd speed parameter
$S.H.$ Specific heat of oil

TABLE 1. *Dimensionless stiffness $\bar{\lambda}_o$ for journal bearings with n pads and without axial slots*

η	Capillary	Orifice	Constant flow	Diaphragm valve
3	$\frac{1{\cdot}08\beta(1-\beta)}{1+1{\cdot}5\gamma(1-\beta)}$	$\frac{2{\cdot}16\beta(1-\beta)}{2-\beta+3\gamma(1-\beta)}$	$\frac{1{\cdot}08\beta}{1+1{\cdot}5\gamma}$	$\frac{1{\cdot}08\beta(1-\beta)}{1+1{\cdot}5\gamma(1-\beta)-\frac{6\beta(1-\beta)}{\bar{\lambda}d}}$
4	$\frac{3{\cdot}82\beta(1-\beta)}{1+\gamma(1-\beta)}$	$\frac{7{\cdot}65\beta(1-\beta)}{2-\beta+2\gamma(1-\beta)}$	$\frac{3{\cdot}82\beta}{1+\gamma}$	$\frac{3{\cdot}82\beta(1-\beta)}{1+\gamma(1-\beta)-\frac{6\beta(1-\beta)}{\bar{\lambda}d}}$
5	$\frac{4{\cdot}12\beta(1-\beta)}{1+0{\cdot}69\gamma(1-\beta)}$	$\frac{8{\cdot}25\beta(1-\beta)}{2-\beta+1{\cdot}38\gamma(1-\beta)}$	$\frac{4{\cdot}25\beta}{1+0{\cdot}69\gamma}$	$\frac{4{\cdot}25\beta(1-\beta)}{1+0{\cdot}69\gamma(1-\beta)-\frac{6\beta(1-\beta)}{\bar{\lambda}d}}$
6	$\frac{4{\cdot}30\beta(1-\beta)}{1+0{\cdot}5\gamma(1-\beta)}$	$\frac{8{\cdot}60\beta(1-\beta)}{2-\beta+\gamma(1-\beta)}$	$\frac{4{\cdot}30\beta}{1+0{\cdot}5\gamma}$	$\frac{4{\cdot}30\beta(1-\beta)}{1+0{\cdot}5\gamma(1-\beta)-\frac{6\beta(1-\beta)}{\bar{\lambda}d}}$

range 5 to 10 times the manufacturing tolerance on h_o. As a guide, choose h_o = 7 times the tolerance.

(7) Specify width of axial flow lands a.
A guide is to calculate $a = b = L/4$ *to* $L/6$.

(8) Specify circumferential flow-land b.
As a guide, $b = \pi D/3n$ (that is, $b = a$ approximately, when $L = Do$.

(9) For an axial exhaust slot bearing, specify c.
As a guide, $c = \pi D/8n$.

(10) Calculate the circumferential flow factor.

(*a*) without axial slots $\gamma = \dfrac{na\,(L-a)}{\pi\,Db}$

(*b*) With axial slots $\gamma = \dfrac{na\,(L-a)}{b\,(\pi D - nc - nb)}$

(11) For bearings with axial slots calculate

$$\theta = \frac{\pi}{n} - \frac{b+c}{D}$$

(12) For bearings which incorporate axial slots, calculate the value for the factor K

$$K = \sin\theta\left[\frac{\sin\theta}{\theta} + \gamma\cos\theta\right]$$

(13) Calculate $\bar{\lambda}_o{}^*$

Use the appropriate equation from Tables 1 or 2 for capillary compensation, and $\beta = 0{\cdot}5$, whatever type of compensation is to be used. This rule is important, since this value is used only as a guide to the maximum safe load of the bearing and is most reliably obtained from the equations for capillary compensation.

(14) Calculate the minimum value of supply pressure P_s.

$$P_s\,(\text{min}) = \frac{3W'}{\bar{\lambda}_o{}^*\,D\,(L-a)}$$

If the value of P_s is too high, refer again to instruc-

TABLE 2. *Dimensionless stiffness $\bar{\lambda}_o$ for journal bearings with n pads and with axial slots*

η	Capillary	Orifice	Constant flow	Diaphragm valve
3	$\frac{4{\cdot}5K\beta(1-\beta)}{1+\gamma(1-\beta)}$	$\frac{9K\beta(1-\beta)}{2-\beta+2\gamma(1-\beta)}$	$\frac{4{\cdot}5K\beta}{1+\gamma}$	$\frac{4{\cdot}5K\beta(1-\beta)}{1+\gamma(1-\beta)-\frac{6\beta(1-\beta)}{\lambda d}}$
4	$\frac{6K\beta(1-\beta)}{1+\gamma(1-\beta)}$	$\frac{12K\beta(1-\beta)}{2-\beta+2\gamma(1-\beta)}$	$\frac{6K\beta}{1+\gamma}$	$\frac{6K\beta(1-\beta)}{1+\gamma(1-\beta)-\frac{6\beta(1-\beta)}{\bar{\lambda}d}}$
5	$\frac{7{\cdot}5K\beta(1-\beta)}{1+\gamma(1-\beta)}$	$\frac{15K\beta(1-\beta)}{2-\beta+2\gamma(1-\beta)}$	$\frac{7{\cdot}5K\beta}{1+\gamma}$	$\frac{7{\cdot}5K\beta(1-\beta)}{1+\gamma(1-\beta)-\frac{6\beta(1-\beta)}{\bar{\lambda}d}}$
6	$\frac{9K\beta(1-\beta)}{1+\gamma(1-\beta)}$	$\frac{18K\beta(1-\beta)}{2-\beta+2\gamma(1-\beta)}$	$\frac{9K\beta}{1+\gamma}$	$\frac{9K\beta\,(1-\beta)}{1+\gamma(1-\beta)-\frac{6\beta(1-\beta)}{\bar{\lambda}d}}$

$$K = \text{Sin}\,\theta\left[\frac{\text{Sin}\,\theta}{\theta\ (\text{radians})} + \gamma\,\text{Cos}\,\theta\right]$$

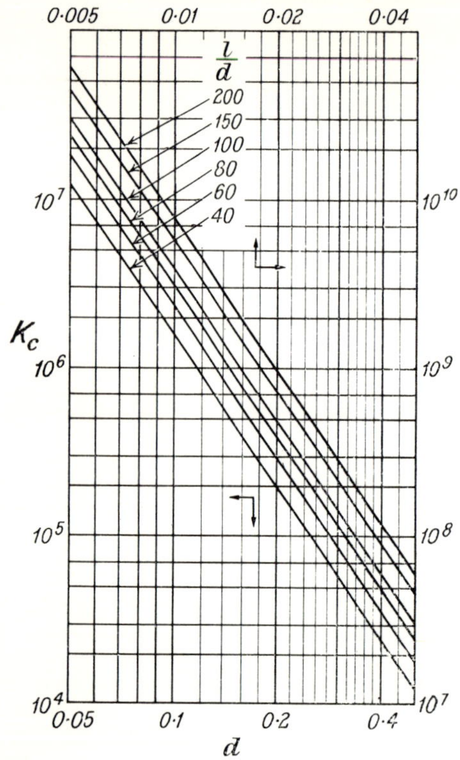

Fig 3

Curves relating to the selection of a capillary bore for a capillary-restrictor type of hydrostatic journal bearing

---

tion (4) or (5), increase D or L (or both) and repeat the calculations.

(15) Decide value of P_s to be used.

(16) Calculate $\bar{\lambda}_o$ from Tables 1 or 2 for appropriate type of compensation.

(17) Calculate stiffness λ_o

$$\lambda_o = \frac{P_s D\,(L - a)}{h_o} \,.\, \bar{\lambda}_o$$

If the stiffness is too low, increase the values of D, L or P_s and repeat.

(18) Calculate the minimum operating gap from the approximate equation

$$h' = h_o - \frac{W'}{\lambda_o}$$

(19) Calculate the flow shape factor $\bar{B}$

(*a*) without axial slots $\bar{B} = \dfrac{\pi D}{6\,an}$

(*b*) with axial slots $\bar{B} = \dfrac{[\pi D - nc - nb]\,[1 + \gamma]}{6\,a\,n}$

(20) Specify the operating speed ν

$$\nu = \frac{\pi D N}{60} \text{ where } N \text{ is speed in r.p.m.}$$

If $\nu = 0$, or is very low as evidenced at instruction (24) by an impracticably high oil viscosity requirement, ignore the effect of speed and specify a convenient value for viscosity.

(21) Calculate the land area for one recess of the bearing only. A_L

(22) Calculate the recess area for one recess of the bearing only. A_R

(23) Calculate the effective friction area for one recess only. A_f

$$A_f = \frac{A_R}{4} + A_L \text{ (reference 5)}$$

(24) Calculate the viscosity. η

The optimum value is given by

$$\eta = \frac{P_s h_o^2}{\nu} \,.\, \sqrt{\left(\frac{\beta \,.\, \bar{B}}{A_f}\right)}$$

and use a consistent set of units. For a system of lbf. inches and seconds η will be in reyns ($\equiv$ lbf. sec. in.$^{-2}$)

If η is impracticably low, first reduce the land widths to a minimum of $100h_o$ and then if necessary increase h_o. Repeat from the appropriate instructions.

(25) Calculate total flow rate. q_o

$$q_o = \frac{P_s h_o^3}{\eta} \,.\, n\beta\bar{B}$$

(26) Calculate the maximum possible temperature rise across the bearing.

(*a*) $\nu \neq 0$

$$\Delta T = \frac{2\,P_s}{\rho \,.\, J \,.\, (S.H.)}$$

for an optimized bearing

(*b*) $\nu = 0$

$$\Delta T = \frac{P_s}{\rho \,.\, J \,.\, (S.H.)}$$

Fig 4

Diagram of a typical orifice restrictor for a hydrostatic journal bearing. The ratio d/D should be less than 0·4 for the design procedure described here

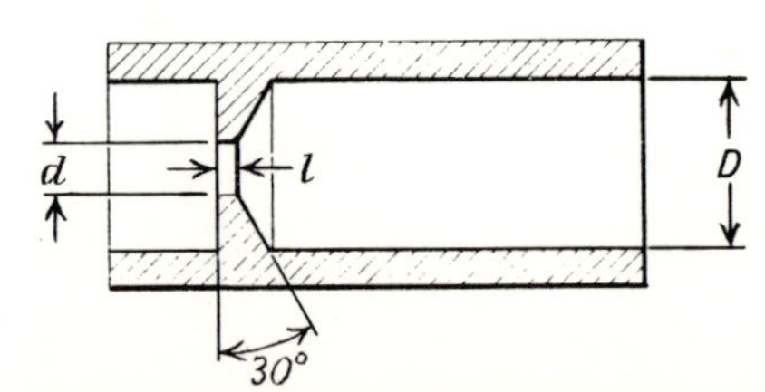

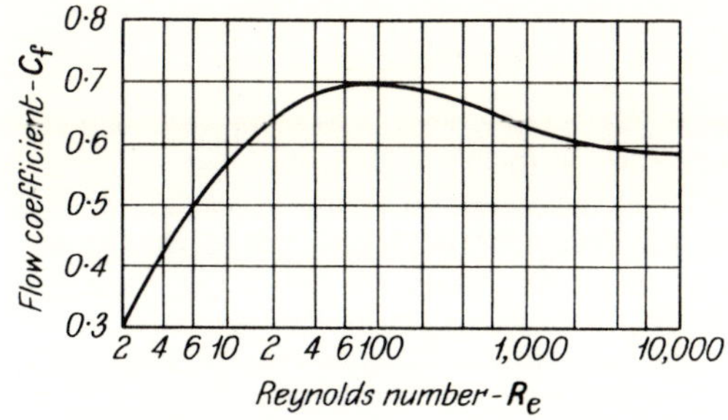

$\rho = 0{\cdot}000078$ to $0{\cdot}000083$ [lbf, in., sec. units]

Fig 5

Curve relating Reynolds number to flow coefficient factor for a range of light-medium machine tool oils

---

It may be noted that when using consistent units in lbf. in. and sec., and for light-medium oils

(*a*) $\nu \neq 0$

$$\Delta T \approx \frac{P_s}{70} \quad \text{(deg. F.)}$$

(*b*) $\nu = 0$

$$\Delta T = \frac{P_s}{140} \quad \text{(deg. F.)}$$

It should be noted that unless the heat is extracted from the oil after passing through the bearing, the total temperature rise will be considerably greater.

Design procedure for capillary restrictors

(1) Determine mass density of oil ρ
In consistent units of lbf. in. and sec.
$\rho = S.G. \times 9{\cdot}33 \times 10^{-5}$ (lbf. in.$^{-4}$ sec.^{2})

(2) Calculate capillary constant K_c

$$K_c = \frac{(1-\beta)\,P_s}{q_o\,\eta}$$

Note: Use q_o value for one recess only.

(3) Choose a value for the capillary bore d (see Fig. 3)

(4) Calculate l

$$l = \frac{\pi d^4\, k_c}{128}$$

l/d should preferably be greater than 100, to eliminate effects of the entrance and exit conditions on the flow.

If l/d is too low, increase d and recalculate l.

(5) Calculate Reynolds number R_e

$$R_e = \frac{4\,\rho\,q_o}{\pi\,d\,\eta}$$

If R_e is greater than 1000, increase d and repeat.

Design procedure for orifice restrictors

A diagram of a typical orifice restrictor for a hydrostatic journal bearing is seen in Fig. 4. Reference should also be made to Fig. 5 and it should be noted that the procedure set out below is based on the assumption that $d/D < 0{\cdot}4$.

(1) Calculate $(q_o)^2$
Note: Use q_o value for one recess only.

(2) Determine mass density of oil ρ
In consistent units of lbf. in. and sec.
$\rho = S.G. \times 9{\cdot}33 \times 10^{-5}$ (lbf. in.$^{-4}$ sec.^{2})

(3) Calculate orifice constant K_{or}

$$K_{or} = \frac{(1-\beta)\,P_s}{q_o^2}$$

Note: Use q_o value for one recess only.

(4) Calculate orifice area A

$$A = \frac{1}{C_f}\cdot\sqrt{\left(\frac{\rho}{K_{or}}\right)}$$

Assume $C_f = 0{\cdot}6$ unless more accurately known.

(5) Calculate orifice diameter d

$$d = \sqrt{\left(\frac{4A}{\pi}\right)}$$

(6) Calculate Reynolds number R_e

$$R_e = \frac{4\,\rho\,q_o}{\pi\,d\,\eta}$$

(7) If necessary correct the value for C_f and repeat (see Fig. 5).

Example—An orifice-compensated journal bearing, without axial exhaust slots, is to support a load of 600 lbf. The shaft rotates at 1800 r.p.m. Determine suitable specifications following the design procedure for a 6-pad bearing.

(1) W	$=$	600 lbf.	
(2) n	$=$	6	
(3) β	$=$	$0{\cdot}5$	
(4) D	$=$	$3{\cdot}0$ in.	$\sqrt{(0{\cdot}015\,W')} = 3{\cdot}0$ in.
(5) L	$=$	$3{\cdot}0$ in.	
(6) h_o	$=$	$0{\cdot}0015$ in.	
(7) a	$=$	$0{\cdot}50$ in.	
(8) b	$=$	$0{\cdot}50$ in.	
(9) c—Inapplicable			
(10) γ	$=$	$\frac{na\,(L-a)}{\pi\,D\,b} = 1{\cdot}6$	
(11) θ—Inapplicable			
(12) K	..		
(13) λ_o*	$=$	$0{\cdot}76$ (From Table 1 for capillary compensation with $\beta = 0{\cdot}5$)	
(14) P_s (min)	$\approx$	300 lbf/in.^{2}	
(15) P_s	$=$	300 lbf./in.^{2}	
(16) λ_o	$=$	$0{\cdot}93$ (From Table 1 for orifice compensation)	
(17) λ_o	$=$	$1{\cdot}4 \times 10^6$ lbf./in.	
(18) h'	$\approx$	$0{\cdot}0011$ in.	

$$(19)\ \bar{B} = \frac{\pi D}{6\,a\,n} = 0{\cdot}52$$

(20) v = 280 in./sec.
(21) A_L = 1·83 in.^{2}
(22) A_R = 2·67 in.^{2}
(23) A_f = 2·50 in.^{2}
(24) η = 0·75 × 10^{-6} reyns
(25) q_o = 2·10 in.^{3}/sec.
(26) Flow = 0·45 gal./min.
(27) ΔT = 4½ deg. F.

It is recommended to make the recess depth 20 times design film thickness.

Consideration at higher speeds

(1) Laminar flow assumption

At higher speeds, the flow may become turbulent and hence invalidate the procedure. The transition speed from laminar to turbulent conditions is yielded from Taylors equation [6] [8].

$$\pi D N_t' h \frac{\rho}{\eta} = 41{\cdot}1 \sqrt{\left(\frac{D}{2h}\right)}$$

where D = journal diameter (in.)
h = radial clearance (in.)
N_t' = transition speed in rev./sec.
ρ = mass density of oil
η = dynamic or absolute viscosity in lbf. sec. in.$^{-2}$ (≡ reyns)

(2) Velocity-induced flow effects

The speed of the bearing induces more flow into the highly-loaded pockets than is exhausted into the lightly-loaded pockets, so that there is always an increase in stiffness at low eccentricity ratios. However, at low values of β it is possible to incur a loss of stiffness at high eccentricity ratios.[9]

Raimondi and Boyd[7] give a speed parameter which is high in the presence of hydrodynamic effects and zero for purely hydrostatic bearings.

$$\omega = \frac{24\pi N' a (L - a)\eta}{h^2 P_s}$$

where ω is the speed parameter
N' is journal speed in rev./sec.
a is axial land width
η is dynamic viscosity
h is radial clearance
P_s is supply pressure

Raimondi and Boyd found that for a typical capillary compensated journal bearing operating at an eccentricity ratio of 0·6, the load capacity was approximately doubled at $\omega = 20$.

(3) Power and temperature rise

At low speeds, power is only dissipated as a result of pumping at supply pressure. At higher speeds, additional energy has to be dissipated due to bearing friction. The total power H_t is the sum of the pumping power H_p and the friction power H_f.

The minimum power condition is always 0 when H_f is in the range

$$H_p \leq H_f \leq 3H_p$$

Because oil is a relatively poor conductor of heat, most of the heat is carried away from the bearing in the oil. As a result it can be shown that the temperature rise as the oil passes through the bearing is directly proportional to supply pressure.

If the optimization is carried out with $H_p = H_f$, the temperature rise is exactly double the temperature rise due to pumping only, and this factor is the basis of the design procedure. The full justification for this approach is given in an article by Rowe and O'Donoghue [4].

Acknowledgements

To *Tribology* for allowing reproduction of various diagrams and Tables from reference [4].

REFERENCES

(1) De Gast: "A New Type of Controlled Restrictor for Double Film Hydrostatic Bearings and its application to High Precision Machine Tools". Proceedings of 7th M.T.D.R. Conference held at Birmingham University 1966 (Pergamon Press) pp 273-298.

(2) W. B. Rowe: "Experience with Four Types of Grinding Machine Spindle". Proceedings of 8th M.T.D.R. Conference, held at Manchester University 1967 (Pergamon Press) pp. 453-476.

(3) H. Opitz: "Externally Pressurized Bearings". Proceedings of Lubrication and Wear Conference, I.Mech.E., Sept. 1967. pp. 67-82.

(4) W. B. Rowe and J. P. O'Donoghue: "Hydrostatic Bearing Design", *Tribology*, Vol. 2, No. 1, February 1969, pp. 25-71.

(5) F. M. Stansfield: "The Design of Hydrostatic Journal Bearings". Proceedings of 8th International M.T.D.R. Conference held at the University of Manchester, Sept. 1967 (Pergamon Press) pp. 419-445.

(6) Wilcock, D. F.: "Turbulence in High Speed Journal Bearings". Trans A.S.M.E., v. 72, 1950, pp. 825-834.

(7) Raimondi and Boyd: "An Analysis of Orifice and Capillary-compensated Hydrostatic Journal Bearings". Lubrication Engineering, January 1967 pp. 28-37.

(8) Fuller, D. D.: "Theory and Practice of Lubrication for Engineers". Wiley.

(9) Davies, P. B. et al: "The Behaviour of a Multi-recess Hydrostatic Journal Bearing". Proceedings of 9th International M.T.D.R. Conference held at Birmingham University Sept. 1968 (Pergamon Press).

* * * * * * * * * * * * * * * * *

Chapter Two

Plane bearings for machine tool slideways

J P O'DONOGHUE and W B ROWE

Hydrostatic bearings are employed in slideway arrangements where the requirements are low friction, low wear, and high accuracy. For many machine tool slideways, the manufacturing costs are not high compared to the alternatives that would be necessary to approach the same standards.

Air support is attractive for many applications and eliminates the requirement for oil collection. However, the maximum load support and dynamic stiffnesses obtainable with this medium are lower than for oil, so that oil lubrication is preferred for machines designed for rapid stock removal.

For square slideways and simple symmetrical arrangements of slides a designer or draughtsman with no previous experience can quickly and reliably arrive at a satisfactory solution. If the problem involves a combination of directional loads and couples acting on a complex arrangement of pads, however, more care must be exercised and reference should be made to a more extensive article by the authors (1)—all references at end of chapter -which deals with all bearing arrangement cases.

Temperature rise is not usually a problem in slideway design. As compared with those for many other bearing applications, the operating speeds are low to moderate, and it is possible to design for low flow-rates. Since heat to be dissipated in a low-speed bearing is proportional to supply pressure and flow-rate it is not usually necessary to make special arrangements for heat exchanging. If the designer is uncertain of the principles of optimization for minimum power and temperature rise, a full discussion may be found in references (2) and (3).

The working principle of a hydrostatic bearing is indicated diagrammatically in Fig. 1, which shows oil from a fixed pressure supply passing through a control device to a bearing pad. In its usual form, the bearing pad contains a pressure equalizing recess surrounded by a restrictive sill or land. The control arrangement is most often a capillary or an orifice, but may also be a constant flow valve or pump. There is also a number of more highly developed arrangements including pressure sensing devices. Control arrangements are discussed fully in references (1) and (4). The control arrangement governs the relationship between bearing thrust W and clearance as follows. For a given bearing thrust there must be a certain pressure in the recess equal to

$$P_r = \frac{W}{Ae}$$

Fig 1

Diagram illustrating the hydrostatic bearing principle

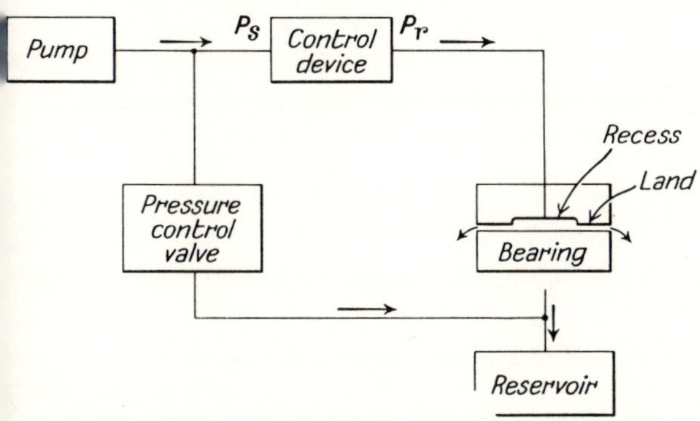

Symbols

P_r	Recess pressure
P_s	Supply pressure
W	Supported load or bearing thrust
A	Total projected bearing area
A_e	Effective area over which P_r may be assumed to act
$\bar{A} = \dfrac{A_e}{A}$	Area shape factor (or load factor) for bearing pad
h	Bearing gap
η	Absolute or dynamic viscosity
$\bar{B} = \dfrac{q\eta}{P_r h^3}$	Flow shape factor for bearing pad
$_o$	Suffix denoting designed operating condition
$\bar{W} = \dfrac{W}{P_s A_e}$	Dimensionless load
λ	Stiffness
$\bar{\lambda} = \dfrac{\lambda h_o}{P_s A_e}$	Dimensionless stiffness
$\bar{X} = h/h_o$	Dimensionless bearing gap
$\beta = \dfrac{P_{ro}}{P_s}$	Pressure ratio for designed operating condition
H	Pumping power
$\bar{H} = \dfrac{HA^2\eta}{Wh^3} = \dfrac{\bar{B}}{\bar{A}^2}$	Dimensionless pumping power
ΔT	Maximum temperature difference from oil inlet to oil outlet (zero speed bearing)

where A_e = effective area of bearing pad. The recess pressure determines the flow-rate q through the control device according to its pressure-flow characteristic. It is the combination of a given flow-rate and recess pressure which determines the bearing gap according to a relationship of form

$$q = \frac{P_r\, h^3}{\eta} \,.\, \bar{B}$$

where h = bearing gap
η = absolute oil viscosity
$\bar{B}$ = flow shape factor

Basic relationships

The designer's primary object is to ensure that

(i) Flow-rate (q) should not be too great
(ii) The bearing should have sufficient stiffness (λ)
(iii) The bearing should support the maximum applied loads without bearing collapse. The load at collapse is W_{ULTIMATE}. In practice it is usual to design for a maximum load which is less than W_{ULTIMATE} and this may be termed W_{MAXIMUM}.

The basic relationships for the calculations may be expressed in terms of the main design parameters, which are supply pressure P_s, design clearance h_o, oil viscosity η, and total projected bearing area A. Calculations which involve the bearing pad geometry and characteristics of the control device may be avoided or simplified by reference to the design graphs for the dimensionless parameters.

The relationships are:—

$$q = \frac{P_r\, h^3}{\eta} \,.\, \bar{B} \qquad \text{(1)}$$

$$W = P_s A_e \,.\, \bar{W}$$
$$= P_s A \,.\, \bar{A}\, \bar{W} \qquad \text{(2)}$$

$$\lambda = \frac{P_s A_e}{h_o} \,.\, \bar{\lambda} \qquad \text{(3)}$$

$$h = h_o \bar{X} \qquad \text{(4)}$$

where $\bar{B}$ is the bearing pad flow factor
$\bar{A}$ is the bearing pad area factor
$\bar{W}$ is the dimensionless thrust
$\bar{\lambda}$ is the dimensionless stiffness
$_o$ is a suffix denoting the designed value
$\bar{X}$ is dimensionless bearing clearance.

Fig 2
A thin land bearing may be of any shape with thin constant land width

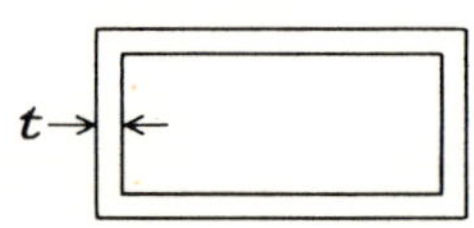

$$A_e = \tfrac{1}{2}[\text{Total area} + \text{recess area}] = A.\bar{A}$$

$$\bar{B} = \frac{[\text{Total area} - \text{recess area}]}{12t^2}$$

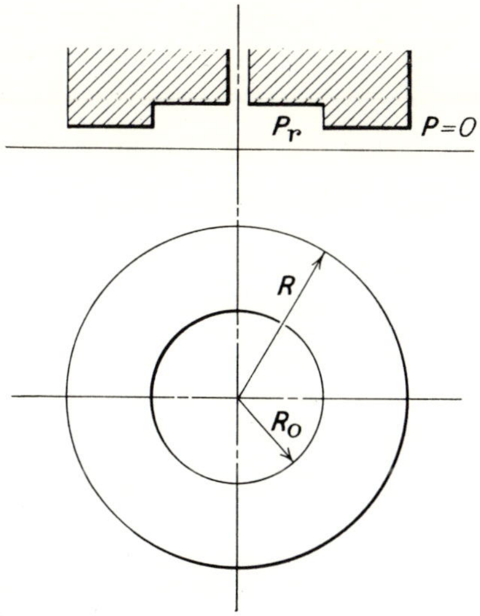

Fig 3

General form of circular thrust pad bearing

Fig 4

Pad coefficients for a circular thrust bearing

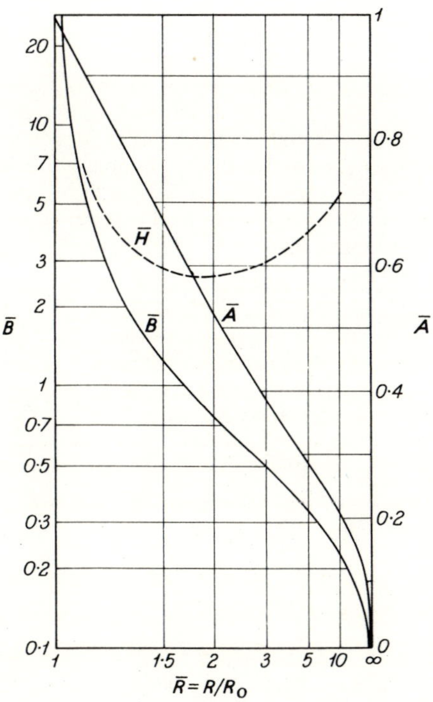

The pad coefficients are $\bar{A}$ and $\bar{B}$, for which there are charts or simple formulae. The relationship between $\bar{W}$, λ and $\bar{X}$ for various control devices may be presented also in chart form. Equation (1) may be written in a form more convenient to the designer for the particular bearing design condition

$$q_o = \frac{P_s h_o^3}{\eta} \cdot \beta \bar{B}$$

where $\beta = \dfrac{P_{ro}}{P_s}$ is the pressure ratio. The pressure ratio β is another important design parameter. Most often the designer chooses $\beta = 0 \cdot 5$, which expresses the intention to adjust the recess pressure to half the supply pressure at the design condition.

Pad coefficients

(i) *Thin land bearings* (Fig. 2)

It may be reasoned [2] that the optimum land width ratio should be 0·25 when the land width is expressed as a proportion of the smaller outside pad dimension. However, thin land bearings are used in two cases:

(*a*) In a given total area to obtain the maximum effective area, and

(*b*) At higher speeds to reduce the area of high friction.

The flow shape factor and effective area for thin land bearings are easily calculated whatever the shape.

$$\bar{B} = \frac{\text{land area}}{12t^2}$$

where t = land width

$$\bar{A} = \frac{\frac{1}{2}\,(\text{Total area} + \text{recess area})}{\text{Total area}}$$

Thin land assumptions may be used as an approximation for flat pads with a land width ratio up to 0·25. The values of $\bar{A}$ and $\bar{B}$ will be over-estimates, but by not more than 7 per cent and 17 per cent respectively of the true values for the worst case.

(ii) *Circular thrust pad* (Fig. 3)

The coefficients for a circular thrust pad have been computed and may be read from Fig. 4. A land width ratio of 0·25 corresponds to $R/R_o = 2$ when $\bar{H}$ is close to its minimum value. $\bar{H}$ is a parameter which expresses the power in a low speed bearing for a given load and a given total projected bearing area. The pumping power is given

$$\text{by} \quad H = P_s q = \frac{Wh^3}{A^2\eta} \cdot \frac{\bar{B}}{\bar{A}^2}$$

$$\text{and} \quad \bar{H} = \frac{\bar{B}}{\bar{A}^2}$$

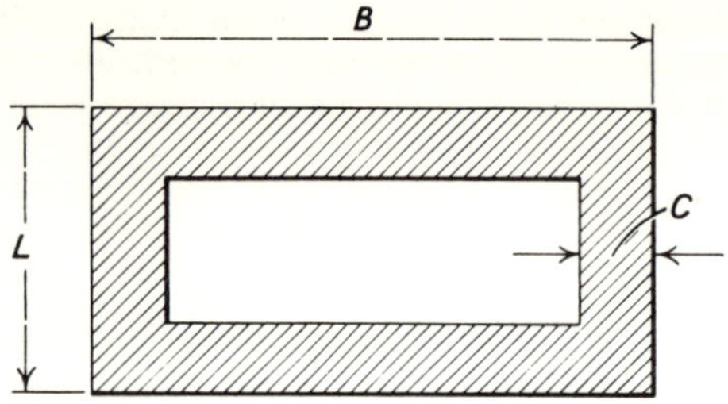

Fig 5

For a rectangular pad with uniform land width it is recommended that $C/L = 0{\cdot}25$

(iii) *Rectangular pads* (Fig. 5)

The pad coefficients (Fig. 6) are presented for various values of C/L and B/L, where the outside dimensions are B and L and the land width is C. It is recommended that C should be $0{\cdot}25 \times L$, which is the shorter outside dimension.

Load, stiffness and gap relationships

The parameters $\bar{W}$, $\bar{\lambda}$ and $\bar{X}$ are independent of pad shapes and depend only on the control device. The relevant equations governing the relationships for single plane bearings are given in Fig. 7.

(1) *Design of single plane bearings* (Fig. 8 to 16)

At the design condition, the maximum stiffness for single plane bearings is obtained when β is approximately $0{\cdot}5$. Increasing β tends to reduce both actual stiffness and actual load range. For this reason values of β greater than $0{\cdot}5$ should be avoided.

If β is reduced, dimensionless stiffness is again reduced. However, if the same applied load is resisted by increasing the bearing area the actual stiffness is increased as is the load range. To reduce the value of β too much causes the stiffness to be much less uniform and leads to an inefficient design system. The preferred range for β is therefore from $0{\cdot}2$ to $0{\cdot}5$ and in certain cases β may be increased up to $0{\cdot}75$.

To obtain a greater load range and higher stiffness there are two possibilities:

(*a*) Increase the bearing area by $2\frac{1}{2}$ times and decrease the design parameter to $0{\cdot}2$.

(*b*) Use opposed pad bearings and greatly increase the bearing area.

The latter recourse yields much higher and more uniform stiffness.

(2) *Design of double plane bearings having equal opposed pads* (Fig. 17)

Double plane bearings may be designed from the data for single plane bearings where

$$Ae = Ae_1 + Ae_2$$
$$\bar{W} = \tfrac{1}{2}(\bar{W}_1 - \bar{W}_2)$$
$$\bar{\lambda} = \tfrac{1}{2}(\bar{\lambda}_1 + \bar{\lambda}_2)$$

With double plane bearings there are fewer situations in which it will be necessary to choose a value of β which is different from $0{\cdot}5$, because load range does not increase greatly with reductions in β and in practice it is more convenient to increase area or supply pressure. Increases in area and supply pressure are made possible since it is not necessary to adjust these values according to the dead weight or design load condition as in single plane bearings.

Fig 6

Coefficients for a rectangular thrust pad with equal land widths

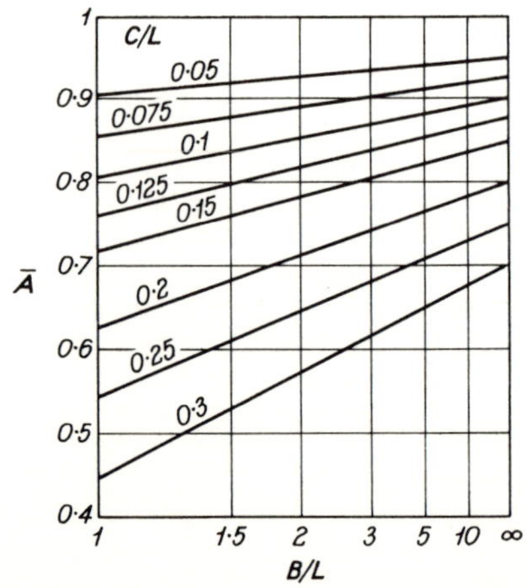

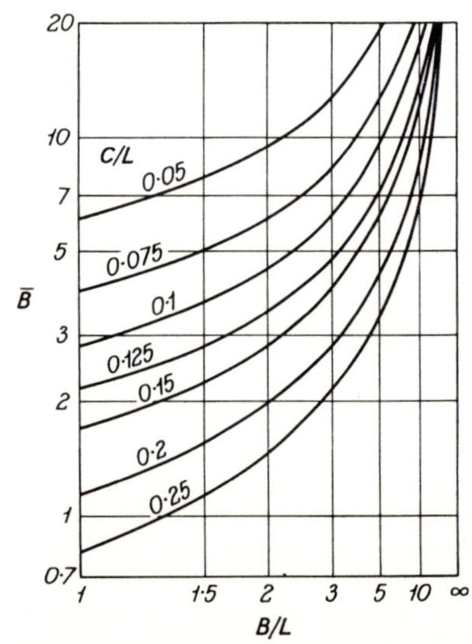

Temperature rise in slow speed bearings

Where temperature rise occurs only due to dissipation of the pumping power, it is a simple matter to calculate the maximum theoretical difference between the temperature of the oil at inlet to the bearing and the temperature at the exit. In practice, however, with massive machine elements and low flow-rates, it is possible to measure a decrease in temperature. This is because the oil tank remains at a higher temperature than the machine.

The maximum difference is calculated assuming that nearly all heat is carried away by the oil so that

$$\Delta T = \frac{P_s}{J \, . \, S \, . \, \rho}$$

where J is the mechanical equivalent of heat
S is the specific heat of the oil
ρ is the mass density

For typical oils used for hydrostatic bearings in machine tools

$$\Delta T = 0{\cdot}0075 \times P_s \ (^\circ\text{F})$$

when P_s is expressed in lbf/in².

EXAMPLE: It is required to support a static load of 1,000 lbf, at zero speed. It has been decided to design an orifice compensated circular pad hydrostatic bearing. Determine suitable specifications, and analyse the bearing performance.

1. Supply pressure. A suitable pressure for many applications is 300 lbf/in². This pressure is high enough to make compressibility effects small and is low enough to ensure that temperature rise is not usually a problem.
2. Pressure ratio β. $\beta = 0{\cdot}5$.
3. Land width ratio (see Fig. 4). Ratio of outer radius to inner radius $R/R_o = 2$.
4. Bearing design clearance h_o. Bearing errors will be a small proportion of the clearance in this case if $h_o = 0{\cdot}0020$ in.
5. Area A. Calculate from $A = \dfrac{W}{P_s \, \beta \, \bar{A}}$ having read $\bar{A} = 0{\cdot}52$ from Fig. 4. Hence
   $$A = \frac{1{,}000}{300 \times 0{\cdot}5 \times 0{\cdot}52} = 14 \text{ in}^2.$$
   Pad dimensions corresponding to $A = 14$ in² and $\bar{R} = 2$, are $R = 2{\cdot}1$ in, and $R_o = 1{\cdot}05$ in.
6. Absolute viscosity η. An oil convenient for many machine tool applications has a viscosity 5×10^{-6} reyns. (The reyn is a unit consistent with lbf, in and sec).
7. Flow rate q. Calculate from
   $$q_o = \frac{P_s h_o^3}{\eta} \, . \, \beta \bar{B},$$ having read from Fig. 4 that
   $\bar{B} = 0{\cdot}75$, hence
   $$q_o = \frac{300 \times 8 \times 10^{-9}}{5 \times 10^{-6}} \times 0{\cdot}5 \times 0{\cdot}75$$
   $$= 0{\cdot}176 \text{ in}^3/\text{sec}.$$
   This state corresponds to $0{\cdot}038$ gal/min.
8. Design stiffness, λ. Calculate from
   $$\lambda_o = \frac{P_s A e}{h_o} \, . \, \bar{\lambda}_o,$$ having read $\bar{\lambda} = 1$ from Fig. 12 for $\bar{X} = 1$.
   $$\text{Hence } \lambda_o = \frac{300 \times 14 \times 0{\cdot}52}{2 \times 10^{-3}} \times 1$$
   $$= 1{\cdot}08 \times 10^6 \text{ lbf/in}.$$
9. Extreme load, W_m. Calculate from $W_m = P_s Ae \, . \, \bar{W}_m$ having examined $\bar{W}$ from Fig. 11, 12 and 13. From Fig. 12, it is seen that stiffness is half the design value at $\bar{X} = 0{\cdot}6$ and $\bar{X} = 1{\cdot}3$.
   From Fig. 11, $\bar{X} = 0{\cdot}6$ corresponds to $\bar{W} = 0{\cdot}95$
   $\bar{X} = 1{\cdot}3$ corresponds to $\bar{W} = 0{\cdot}25$
   $$\text{Hence } W_{\text{MAX}} = 300 \times 14 \times 0{\cdot}52 \times 0{\cdot}95 = 2{,}060 \text{ lbf}$$
   $$W_{\text{MIN}} = 300 \times 14 \times 0{\cdot}52 \times 0{\cdot}25 = 543 \text{ lbf}$$
10. Clearance when $W = 1{,}600$ lbf.
    First calculate
    $$\bar{W} = \frac{W}{P_s A_e} = \frac{1{,}600}{300 \times 14 \times 0{\cdot}52} = 0{\cdot}735$$
    From Fig. 11, $\bar{X} = 0{\cdot}75$ and
    $$h = 0{\cdot}75 \times 0{\cdot}002 = 0{\cdot}0015 \text{ in}$$

Fig 7
Thrust, stiffness and gap equations for single plane bearings

CAPILLARY	ORIFICE	CONSTANT FLOW
$\bar{P}_r = \dfrac{1}{1 + \dfrac{1-\beta}{\beta} \cdot \bar{X}^3}$	$\bar{X}^6 = \dfrac{1-\bar{P}_r}{\bar{P}_r^2} \cdot \dfrac{\beta^2}{1-\beta}$	$\bar{P}_r = \dfrac{\beta}{\bar{X}^3}$ N.B. $\bar{P}_r \ngtr 1$
$\bar{\lambda} = \dfrac{3\bar{X}^2 . \dfrac{\beta}{1-\beta}}{\left[\bar{X}^3 + \dfrac{\beta}{1-\beta}\right]^2}$		$\bar{\lambda} = \dfrac{3\beta}{\bar{X}^4}$ $\lambda = \dfrac{A_e}{h} \cdot 3P_r$
$\bar{\lambda}\bar{X} = 3\bar{P}_r(1-\bar{P}_r)$	$\bar{\lambda}\bar{X} = 6\bar{P}_r\left[\dfrac{1-\bar{P}_r}{2-\bar{P}_r}\right]$	$\bar{\lambda}\bar{X} = 3\bar{P}_r$

8

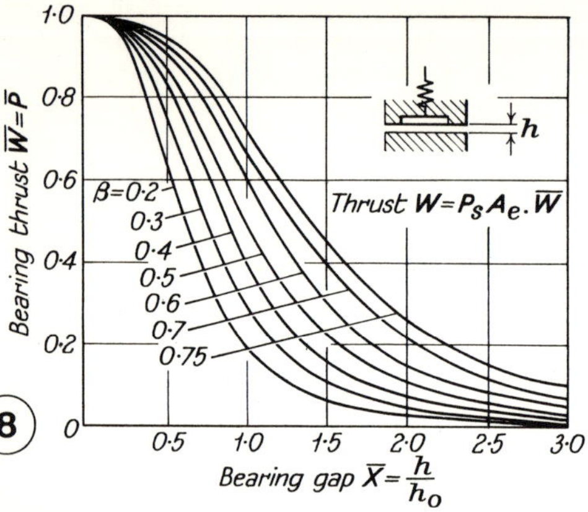

11

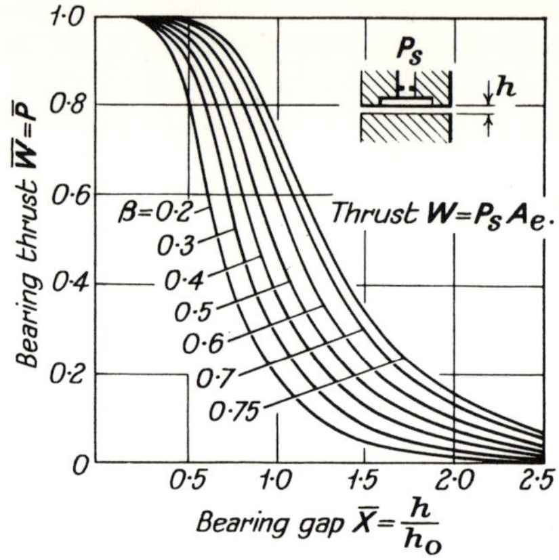

9

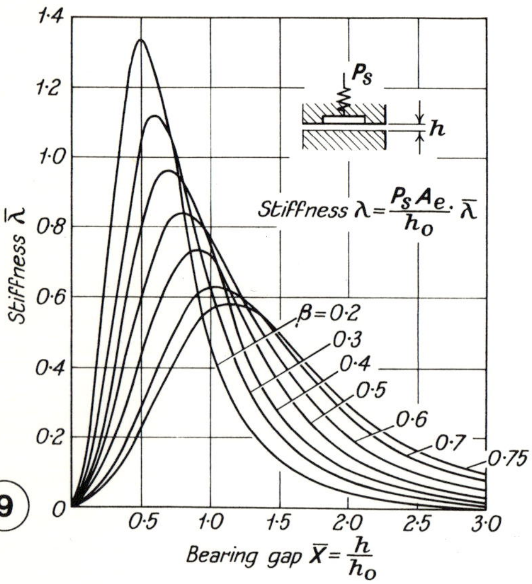

12

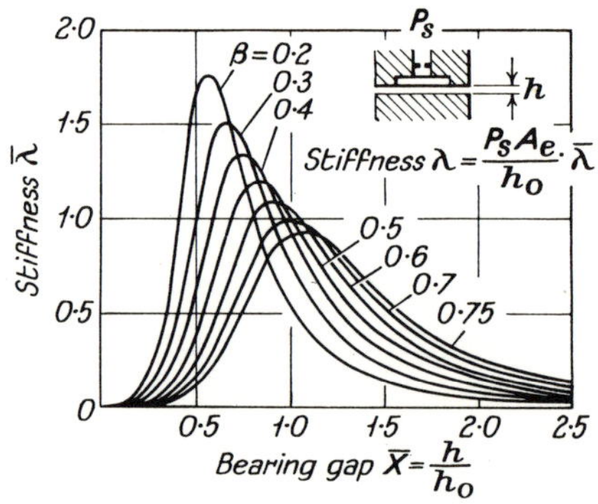

13

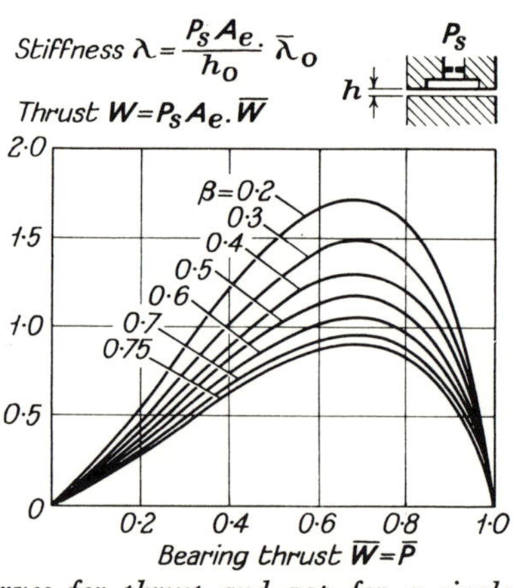

10

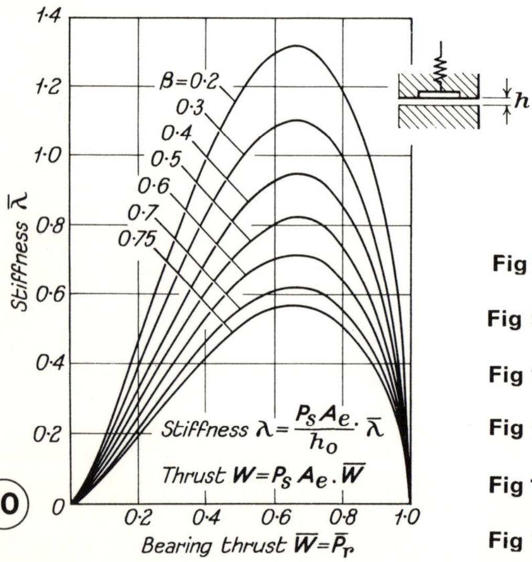

Fig 8 *Curves for thrust and gap for a single plane bearing with capillary control*

Fig 9 *Curves for stiffness and gap, for a single plane bearing with capillary control*

Fig 10 *Curves for stiffness and thrust, for a single plane bearing with capillary control*

Fig 11 *Curves for thrust and gap, for a single plane bearing with orifice control*

Fig 12 *Curves for stiffness and gap, for a single plane bearing with orifice control*

Fig 13 *Curves for stiffness and thrust, for a single plane bearing with orifice control*

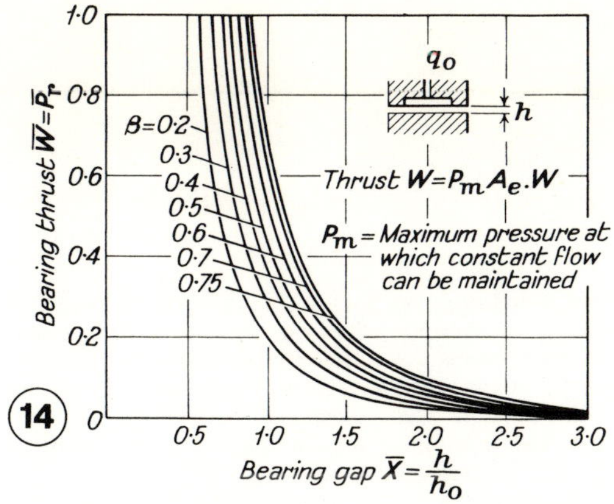

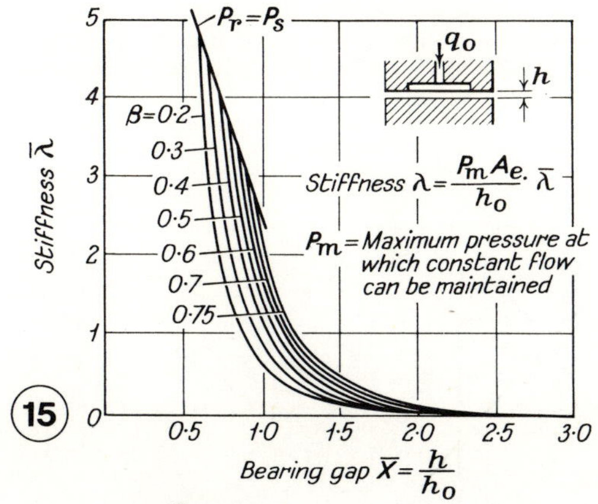

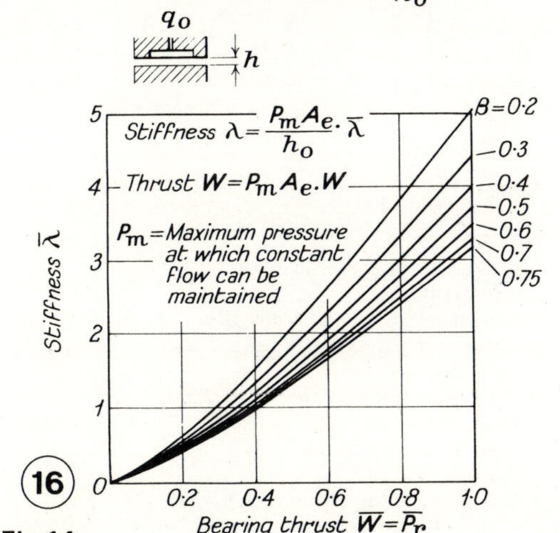

Fig 14
Curves for thrust and gap, for a single plane bearing with constant flow

Fig 15
Curves for stiffness and gap, for a single plane bearing with constant flow

11. Stiffness, when $W = 1{,}600$ lbf.
From Fig. 13, $\bar{\lambda} = 1{\cdot}2$

$$\lambda = \frac{P_s A_e}{h_o} . \bar{\lambda} = \frac{300 \times 14 \times 0{\cdot}52}{2 \times 10^{-3}} \times 1{\cdot}2$$
$$= 1{\cdot}3 \times 10^6 \text{ lbf/in}$$

12. Maximum temperature difference ΔT
Calculate from $\Delta T = 0{\cdot}0075\ P_s$
$T = 0{\cdot}0075 \times 300 = 2\tfrac{1}{4}°$ F.

Acknowledgements are due to *Tribology* for permission to reproduce diagrams from reference (1).

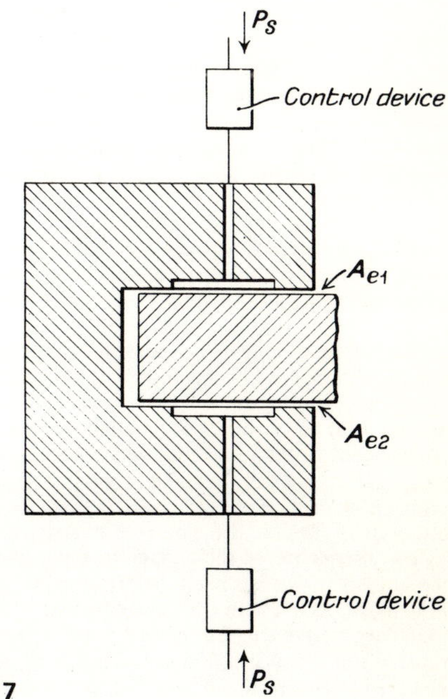

Fig 17
Double plane bearing with equal opposed pads

REFERENCES

1. J. P. O'Donoghue and W. B. Rowe: 'Hydrostatic Bearing Design', *Tribology*, Vol. 2, No. 1, February, 1969, pp. 25-71.
2. W. B. Rowe, J. P. O'Donoghue and A. Cameron: 'Optimisation of Externally Pressurised Bearings for Minimum Power and Low Temperature Rise', *Tribology*, Vol. 4, August, 1970. pp. 153-157.
3. J. P. O'Donoghue and W. B. Rowe: 'Optimisation of Hydrostatic Bearings for Minimum Power and Low Temperature Rise', *Industrial and Scientific Lubrication*, April, 1969, pp. 101-106.
4. J. P. O'Donoghue and W. B. Rowe: 'Compensation Methods for Externally Pressurised Bearings', *Engineers Digest*, April, 1969, Vol. 30, No. 4, pp. 49-61.

Fig 16
Curves for stiffness and thrust for a single plane bearing with constant flow

Chapter Three

Conical journal bearings

R L ASTON, J P O'DONOGHUE and W B ROWE

The first two chapters in this book dealt with cylindrical journal bearings and plane pad bearings for slideways.

More extensive information on the subject may be found in references (1, 2). There are also computer-aided design procedures available, details of which may be obtained on request from the authors.

A conical arrangement for hydrostatic journal bearings may be an advantage for two reasons. Firstly, this arrangement eliminates the need for separate thrust and journal bearings, and secondly, clearance is easily adjustable on assembly.

The basic principle of a hydrostatic bearing applied to a conical arrangement is shown in Fig 1. Each pocket is supplied from a constant pressure source by way of a control arrangement. It is necessary to have a separate control arrangement (usually a capillary) for each pocket. Under load, the clearance in some pockets tends to decrease thus providing increased constriction to flow. The pressure drop across the control arrangement is reduced as a consequence of the change in flow, and the pocket pressure increases to react to the applied load. The opposite effect occurs with the lightly-loaded pockets.

Four possible arrangements for conical journal bearings are illustrated in Fig 2. It is better that the bearings should be well separated in the axial direction rather than closely spaced. The effect of close spacing under combined axial and radial loadings, as in Fig 2a, may be assessed by considering the resulting bearing clearances. Axial load will displace the shaft axially so that clearance in one cone is reduced. A radial displacement of the shaft will reduce the minimum film thickness of each bearing by the same amount. The maximum radial displacement is governed by the small-clearance cone and at this condition the small-clearance cone may carry an appreciable radial load. However, the large-clearance cone does not carry any appreciable radial load. The advantage of the widely spaced arrangements is that the shaft may tilt slightly, thus allowing the large clearance cone to assume a larger operating eccentricity. Both bearings may contribute a

Fig 1

Hydrostatic principle applied to conical journal bearings

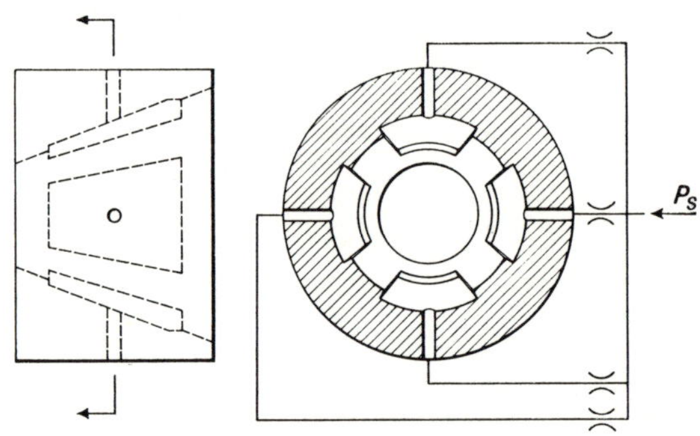

Symbols

P_s Supply pressure
P_o Recess pressure at design condition
h_o Normal clearance (see Fig 4)
C_D Equivalent diametral clearance $= 2h_o$
D Bearing diameter at large end
L Bearing length along axis
a Axial flow land width along axis
b Circumferential flow land width at large end
α Half cone-angle
β Design pressure ratio $= P_o/P_s$
η Viscosity
T Maximum axial load
$\bar{W}$ Dimensionless radial load $= W/P_sLD$
$\bar{T}$ Dimensionless axial load $= T/P_sD^2$
λ_r Radial stiffness of one bearing
λ_a Axial stiffness
$\bar{\lambda}_r$ Dimensionless radial stiffness $= \lambda_r h_o/P_sLD$
$\bar{\lambda}_a$ Dimensionless axial stiffness $= \lambda_a h_o/P_sD^2$
q Flow-rate
$\bar{Q}$ Dimensionless flow-rate $= q\eta/P_s h_o^3 \, . \, a/D$
ρ Mass density of oil
$S.G.$ Specific gravity of oil
C Specific heat of oil
ΔT Maximum temperature difference between oil inlet and oil outlet
d Capillary diameter
l Capillary length
K Capillary constant $= 128l/\pi d^4$
J Mechanical equivalent of heat
R_e Reynolds number

large thrust and hence the total radial load supported is almost doubled.

The closely spaced configuration of Fig 2b is even worse than the previous case. Since the section of the four pads is disposed around a circle there will be poor resistance to shaft tilt about an axis through the circle centre. The widely spaced configuration of Fig 2c is an improvement but from the previous considerations it will be apparent that the best configuration is the widely spaced "back to back" arrangement as shown in Fig 2d. When the loads are applied at one end of the spindle instead of centrally, it is permissible to reduce the radial projected area of one bearing as illustrated in Fig 3. This arrangement has the advantage that power dissipation is reduced. The axial projected areas should be approximately equal.

The design procedure described below is intended primarily for centrally-loaded bearings, but may be used safely for widely-spaced configurations with overhung loads. The calculation example given applies to such a case.

Bearing geometry

The symbols used to describe a bearing are indicated in Fig 4. For slow speed bearings, the land width ratio a/L should be 0·25. At higher speeds this ratio may have to be reduced as indicated in the design procedure. The width of the lands separating the pockets should be in proportion to the diameter. It is recommended that the two land width ratios are made equal, i.e. $a/L = b/D$, which is the condition assumed for computing the charts.

Pressure ratio

The capillaries feeding each pocket should be of such a length that the pocket pressures in the lightly-loaded condition are all 0·3 times the supply pressure, i.e. $\beta = 0{\cdot}3$. A design procedure for capillaries follows the procedure for designing the bearings, which is the usual sequence of calculation.

Design charts

The design charts, Fig 5-9, are based on capillary control and enable bearings to be designed with various length/diameter ratios and cone angles. Fig 5 gives an indication of suitable combinations of length/diameter ratio and cone angle, and Fig 6, 7, 8 are computed values of load capacities, stiffnesses, and flow-rates referred to in the design procedure.

Consideration at higher speeds

At higher speeds, the same considerations apply as described previously for cylindrical bearings 1 . These considerations may be stated briefly as follows:

(*a*) At very high speeds there is a risk of turbulence in the pockets so that the basic assumption of laminar flow is no longer valid.

(*b*) If the bearings are designed according to the following procedure for minimum power, then hydrodynamic effects will not dominate bearing performance. However, if this aspect is ignored, the bearing may be inefficient and may also cavitate with a consequent reduction in dynamic stiffness.

(*c*) The procedure implicitly optimises the design for minimum power and low temperature rise on the basis that pumping power must equal friction power. It is important that this aspect is given some attention, firstly

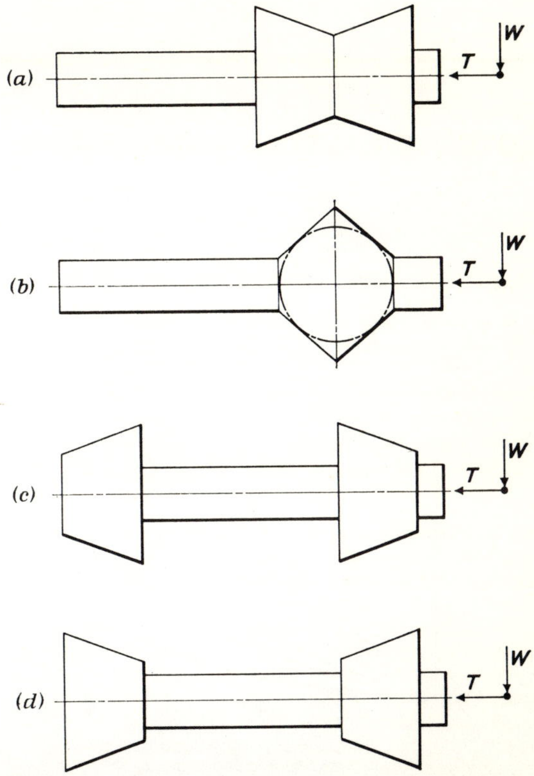

Fig 2

Comparison of four bearing arrangements. (a) Not recommended because of reduced capacity under combined axial and radial loads. (b) Very poor-reduced capacity under combined loads and low resistance to tilt. (c) Improved bearing arrangement. (d) Best bearing arrangement

Fig 3

Modification of bearing arrangement for loads applied mainly at ane end. (a) Configuration for centrally applied loads. (b)Modified configuration for loads applied mainly at one end. Equal areas for axial loads. Unequal areas for radial loads

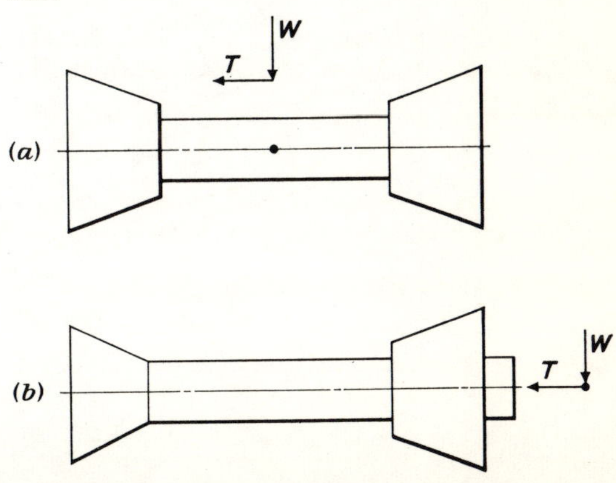

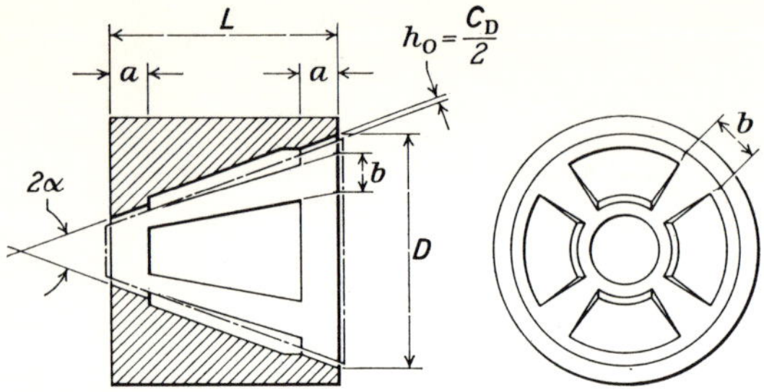

Fig 4

Description of conical hydrostatic bearing geometry

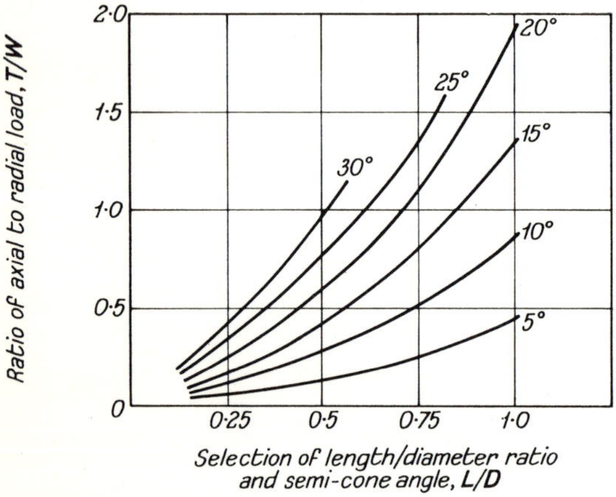

Fig 5

Curves for selecting length/diameter ratio and semi-cone angle

in order to avoid cavitation as mentioned above and secondly in order to avoid excessive heat exchanging requirements.

The instruction which ensures minimum power is the calculation of viscosity from

$$\eta = \frac{P_s}{N'} \cdot \left(\frac{C_D}{D}\right)^2 . S_H$$

If on making this calculation the value of η is impracticably low for the application, as a result of the high speed, it may be necessary to readjust the pocket lands and the clearance. The easiest way to obtain the 'next best' bearing for a given viscosity is as follows. Reduce the lands until $a = 50h_o$ and calculate a new value of diametral clearance from

$$C_D = D\sqrt{\frac{\eta N}{S_H P_s}}$$

A conveniently high value of viscosity may be adopted for a bearing which operates at low speed. It will be evident that the bearing is low speed when the calculated value of viscosity appears impracticably high.

The necessary values of S_H are given in Fig 9.

Design procedure for opposed conical bearings

1. Radial load W (lbf)
   Specify maximum load to be carried by *main* bearing
2. Axial thrust load T (lbf)
   Specify maximum axial load to be carried
3. Operating speed N' (rev/sec)
4. Supply pressure P_s (lbf/in²)
   Decide a suitable value. 300 lbf/in² is a convenient value for many pumps.
5. Length/diameter ratio L/D (—)
   Select a suitable combination of L/D and α from Fig 5 for T/W equal to or less than required value.

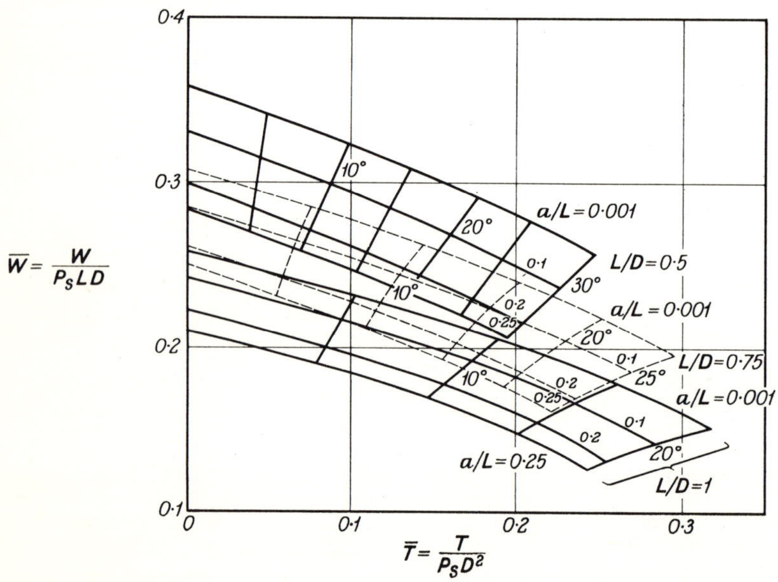

Fig 6

Radial and axial loads that may be safely applied in combination

6. Semi-cone angle α (—)
Fig 5.
7. Land-width ratio a/L (—)
Decide $a/L = 0{\cdot}25$ initially and $b/D = a/L$
8. Dimensionless thrust $\overline{T}$ (—)
Read from Fig 6
9. Diameter D (in)
Calculate from $D = \sqrt{\dfrac{T}{P_s \overline{T}}}$
10. Length L (in)
From $D \times L/D$
11. Clearance h_o (in)
Specify normal clearance h_o approximately $5 \times$ manufacturing tolerance on h_o
$$C_D = 2 \times h_o$$
12. Optimization parameter S_H (—)
Read from Fig 9. If $N' = 0$ specify convenient value of η and continue from instruction 18.
13. Absolute viscosity η (reyns = lbf sec in^{-2})
Calculate from $\eta = \dfrac{S_H P_s}{N'} \left(\dfrac{C_D}{D}\right)^2$
Continue from instruction 18 unless calculated value of η is too low, in which case specify minimum acceptable value. If calculated value of η is too high, specify a convenient value and continue from instruction 18.
14. S_H for minimum viscosity (—)
Calculate new value from $S_H = \dfrac{N' \eta}{P_s} \left(\dfrac{D}{C_D}\right)^2$
15. Land width ratio (new value) $a/L = b/D$ (—)
Read from Fig 9 for new value of S_H. Continue from instruction 18 unless $a < 50h_o$ in which case specify $a = 50h_o$
16. S_H for minimum land widths (—)
Read from Fig 9 for new value of a/L
17. Clearance (new value) h_o (in)
Calculate from
$$h_o = \frac{C_D}{2} = \frac{D}{2}\sqrt{\frac{\eta N'}{S_H P_s}}$$
18. Dimensionless flow $\overline{Q}$ (—)
Read from Fig 8
19. Flow rate q (in^3/sec)
Calculate from
$$q = \frac{2 P_s h_o^3}{\eta} \cdot \frac{1}{L/D} \cdot \frac{1}{a/L} \cdot \overline{Q}$$
The result is the total flow for two bearings. Multiply the value by 60/277 to obtain the flow rate in gall/min.
20. Dimensionless radial stiffness $\bar{\lambda}_r$ (—)
Read from Fig 7.
21. Radial stiffness λ_r (lbf/in)
Calculate from
$$\lambda_r = \frac{P_s L D}{h_o} \bar{\lambda}_r$$
The result is the radial stiffness of each bearing.
22. Dimensionless axial stiffness $\bar{\lambda}_a$ (—)
Read from Fig 7.
23. Axial stiffness λ_a (lbf/in)
Calculate from
$$\lambda_a = \frac{P_s D^2}{h_o} . \bar{\lambda}_a$$
24. Maximum power dissipation H_t (lbf/in/sec)
If $N' = 0$ (or is very low) $H_t = P_s q$
$N' \neq 0$ $H_t = 2P_s q$
25. Maximum temperature difference ΔT (deg F)
If $N' = 0$ (or is very low) $\Delta T = 0{\cdot}0075\, P_s$
$N' \neq 0$ $\Delta T = 0{\cdot}015\, P_s$ working in a system of in, lbf and sec.

Design procedure for capillary restrictors

1. Determine mass density of oil ρ in consistent units of lbf, in and sec.
$\rho = S.G. \times 9{\cdot}33 \times 10^{-5}$ (lbf. in^{-4} sec^2)
2. Calculate capillary constant K_c based on flow q

Fig 7
Design chart for average radial and axial stiffness

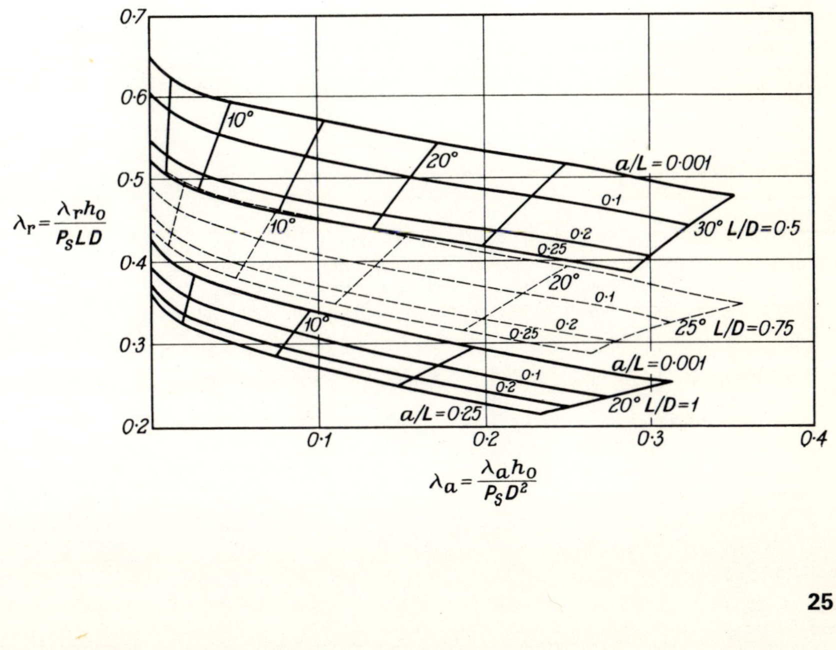

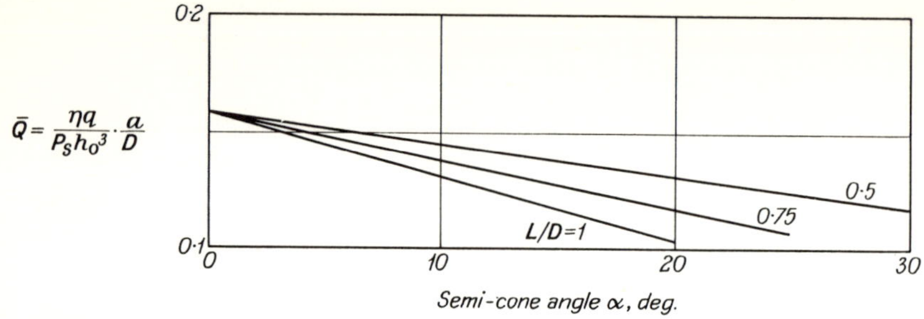

Fig 8
Design chart giving computed values for flow rate

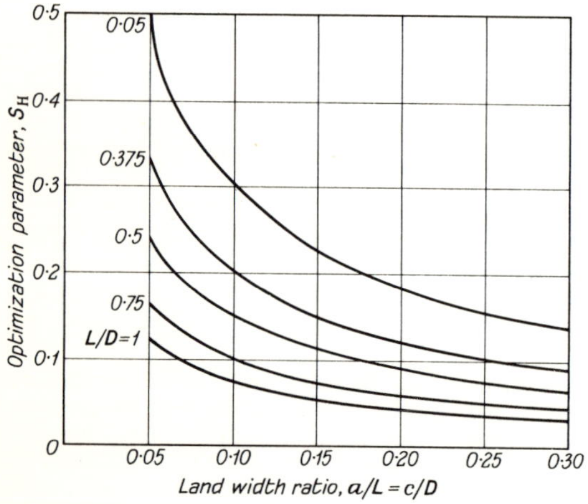

Fig 9

Design chart giving optimization parameter values for conical and cylindrical journal arrangements with pressure ratio $\beta = 0 \cdot 3$

through one recess only and $\beta = 0 \cdot 3$

$$K_c = \frac{(1 - \beta)\, P_s}{q\, \eta} \qquad (\text{in}^{-3})$$

3. Calculate capillary bore d from $d^3 = 128\, l/k_c d$ for a suitable value of l/d
4. Read l (in)
   l/d should preferably be greater than 100, to eliminate effects of the entrance and exit conditions on the flow.
5. Calculate Reynolds Number R_e

   $$R_e = \frac{4\, \rho\, q}{\pi\, d\, \eta}$$

   If R_e is greater than 1,000, increase d and repeat.

Example

A machine spindle is to withstand 300 lbf in the axial direction and 500 lbf radially on the main bearing. The operating speed is 1,200 rev/min. The minimum viscosity is $1 \cdot 2 \times 10^{-6}$ reyns at the working temperature. The design pressure ratio is to be 0·3.

1. $W = 500$ lbf
2. $T = 300$ lbf
3. $N' = 1{,}200/60 = 20$ rev/sec
4. $P_s = 300$ lbf/in²
5. $L/D = 0 \cdot 5$
6. $\alpha = 20°$
7. $a/L = b/D = 0 \cdot 25$ initially
8. $\bar{T} = 0 \cdot 135$
9. $D = \sqrt{\dfrac{300}{300 \times 0 \cdot 135}} = 2 \cdot 7$ in
10. $L = 0 \cdot 5 \times 2 \cdot 7 = 1 \cdot 35$ in
11. $h_o = 0 \cdot 001$ in initially, $C_D = 0 \cdot 002$ in
12. $S_H = 0 \cdot 08$
13. Calculated $\eta = \dfrac{0 \cdot 08 \times 300}{20} \left(\dfrac{0 \cdot 002}{2 \cdot 7}\right)^2$
    $= 0 \cdot 66 \times 10^{-6}$ reyns
    Minimum $\eta = 1 \cdot 2 \times 10^{-6}$ reyns
14. $S_H = \dfrac{1 \cdot 2 \times 10^{-6} \times 20}{300} \left(\dfrac{2 \cdot 7}{0 \cdot 002}\right)^2 = 0 \cdot 15$
15. $a/L = 0 \cdot 1$. This increases $\bar{T}$ and $\bar{W}$ and may be used to reduce D if required.
16. —
17. —
18. $\bar{Q} = 0 \cdot 132$
19. $q = 2 \times \dfrac{300 \times (0 \cdot 001)^3 \times 0 \cdot 132}{1 \cdot 2 \times 10^{-6} \times 0 \cdot 1 \times 0 \cdot 5} = 1 \cdot 32$ in³/sec
    $\equiv 0 \cdot 29$ gal/min
20. $\bar{\lambda}_r = 0 \cdot 5$
21. $\lambda_r = \dfrac{300 \times 1 \cdot 35 \times 2 \cdot 7 \times 0 \cdot 5}{0 \cdot 001} = 0 \cdot 55 \times 10^6$ lbf/in
22. $\bar{\lambda}_a = 0 \cdot 16$
23. $\lambda_a = \dfrac{300 \times (2 \cdot 7)^2}{0 \cdot 001} \times 0 \cdot 16 = 0 \cdot 35 \times 10^6$ lbf/in
24. $H_t = 2 \times 300 \times 1 \cdot 32 = 792$ lbf in/sec $\equiv 0 \cdot 12$ h.p.
25. $\Delta T = 0 \cdot 015 \times 300 = 4\tfrac{1}{2}$°F

To design capillaries for flow rate $q = 1 \cdot 32/8 = 0 \cdot 165$ in³/sec and $S.G. = 0 \cdot 87$

1. $\rho = 0 \cdot 87 \times 9 \cdot 33 \times 10^{-5} = 8 \cdot 1 \times 10^{-5}$ lbf in⁻⁴ sec²
2. $k_c = \dfrac{(1 - 0 \cdot 3)\, 300}{0 \cdot 165 \times 1 \cdot 2 \times 10^{-6}} = 1 \cdot 06 \times 10^{+9}$ in⁻³
3. $d^3 = \dfrac{128 \times 100}{\pi \times 1 \cdot 06 \times 10^{+9}}$ for $l/d = 100$
   $= 0 \cdot 38 \times 10^{-3}$ in³
   $d = 0 \cdot 073$ in
4. $l = 7 \cdot 3$ in
5. $R_e = \dfrac{4 \times 8 \cdot 1 \times 10^{-5} \times 0 \cdot 165}{\pi \times 0 \cdot 073 \times 1 \cdot 2 \times 10^{-6}} = 196$

   This value is sufficiently low and therefore the design is satisfactory.

REFERENCES

1. J. P. O'Donoghue and W. B. Rowe, "Hydrostatic Bearing Design," *Tribology*, Vol. 2, No. 1, Feb. 1969, pp. 25-71.
2. J. P. O'Donoghue and W. B. Rowe, "Hydrostatic Journal Bearing (Exact Procedure)," *Tribology*, Vol. 1, November 1968, pp. 230-236.

Chapter Four

Opposed-pad bearings for machine tool slideways

J P O'DONOGHUE and W B ROWE

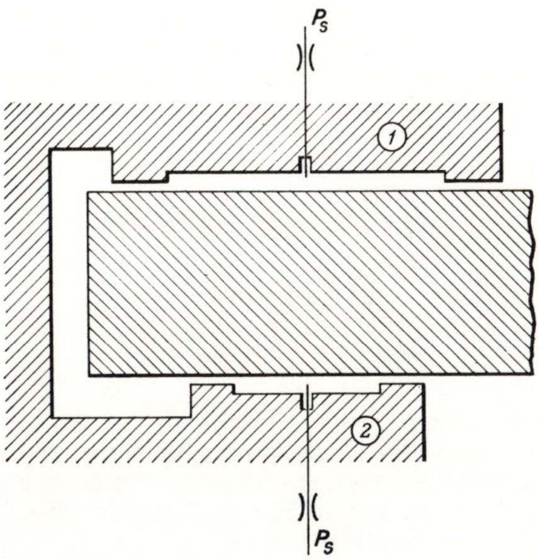

Fig 1

General form of opposed-pad bearing

In many applications of hydrostatic bearings to machine tools and other machinery, opposed pad configurations are employed. As in the previous design procedure for journal bearings , plane pad bearings , and conical bearings , the aim has been to simplify the calculations as far as possible. The following data are sufficient to allow a designer to arrive at a satisfactory design very quickly. For designers who wish to study the subject of hydrostatic bearing design in greater depth, reference may be made to other articles by the same authors[1, 4].

In addition to design procedures from charts, "on-line" computer-aided design techniques for optimized hydrostatic bearings are available which allow the calculations to be completed very quickly and reliably and require even less familiarity with mathematics than the procedure outlined in this chapter. Further details on this aspect may be obtained from the authors.

Opposed pad bearings

Opposed-pad bearings are used where it is necessary to allow for force reversals. The pads may be equal or unequal depending on the relative size of the forces. Unequal opposed pads are used if the principal forces occur in one direction but allowance must be made for smaller forces in the reverse direction. Opposed pad arrangements have the advantage of the total stiffness of both pads but have a load capacity less than that of either

Symbols

Suffix f Denotes friction as in effective area Af
Suffix o Denotes design condition
Suffix r Denotes condition at bearing recess
Suffix s Denotes supply
Suffix 1, 2 Denotes first (large) bearing pad and second (small) bearing pad respectively
P_r Recess pressure
P_s Supply pressure
W Supported load
A Total projected bearing area
A_e Effective area over which P_e may be assumed to act
$E = \dfrac{A_{e1}}{A_{e2}}$ Area ratio
$\bar{A} = \dfrac{A_e}{A}$ Area shape factor (or load factor) for bearing pad
C Sill (or land) width
D Circular pad diameter
L Rectangular pad length (short side)
h Bearing gap
η Absolute or dynamic viscosity
$\bar{B} = \dfrac{q\eta}{P_s h^3}$ Flow shape factor for bearing pad
$\bar{W} = \dfrac{W}{P_s(A_{e1} + A_{e2})}$ Dimensionless load
λ Stiffness
$\bar{\lambda} = \dfrac{\lambda h_o}{P_s(A_{e1} + A_{e2})}$ Dimensionless stiffness
$\bar{X} = h/h_o$ Dimensionless gap
$\beta = P_{ro}/P_s$ Pressure ratio for designed operating condition
H Pumping power
$\bar{H} = \dfrac{HA^2\eta}{Wh^3} = \dfrac{\bar{B}}{\bar{A}^2}$ Dimensionless pumping power
ΔT Maximum temperature difference from oil inlet to oil outlet (zero speed bearing)

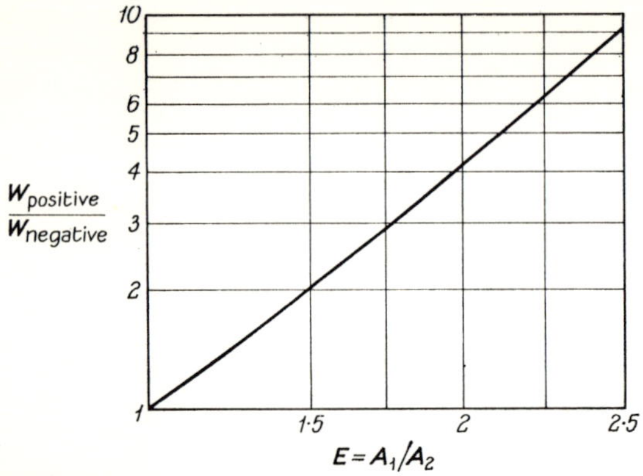

Fig 2

Graph for selection of area ratio for opposed-pad bearings

Fig 3

General form of circular thrust pad bearing

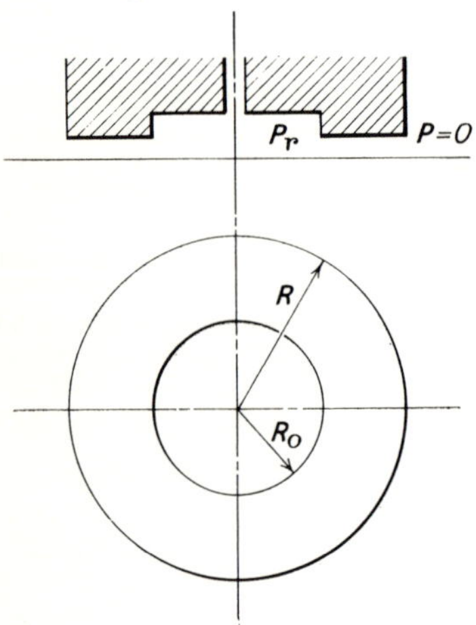

of the pads. Since machine tool slideways have generally low traversing speeds, the problems of high speed operation are not dealt with in this chapter but further consideration of this effect may be found in reference[1].

The basic arrangement of the bearing is shown in Fig 1. The two pads are fed from a common pressurized supply through a control arrangement. The design charts given in this chapter are for a pressure ratio of 0·5, i.e. $\beta = P_o/P_s = 0{\cdot}5$. This value of β is a useful compromise for good load and stiffness, as explained in reference[1]. More detailed charts are contained in reference[3] for unequal opposed pads, the charts here being simplified for design use.

Basic relationships for load, stiffness and flowrate

The load and stiffness relationships required by the designer are given on the charts, and are as follows.

$$W = P_s\,(A_{e1} + A_{e2})\,\bar{W} \text{ for bearing thrust} \quad \ldots\ldots \quad (1)$$

and

$$\lambda = \frac{P_s(A_{e1} + A_{e2})}{h_o}\,\bar{\lambda} \text{ for bearing stiffness} \quad \ldots\ldots \quad (2)$$

The equation for flowrate at the design condition is

$$q_o = \frac{P_s h_o^3 \,.\, 2\beta\bar{B}}{\eta} \quad \ldots\ldots\ldots\ldots \quad (3)$$

In order to simplify the design procedure as far as possible equations (1) and (2) have been written in the form

$$W = P_s A\bar{A} \,.\, \bar{W} \quad \ldots\ldots\ldots\ldots \quad (4)$$

$$\lambda = \frac{P_s A\bar{A}}{h_o} \,.\, \bar{\lambda} \quad \ldots\ldots\ldots\ldots \quad (5)$$

where $\bar{A}$ is a measure of the extent to which recess pressure may be assumed to act on the pad area. These equations are based on the assumption that the proportions of the two pads will be the same and only the size will be different, so that the pad shape factors $\bar{A}$ are identical for the two pads. The same applies for the flow shape factors $\bar{B}$. The values of $\bar{A}$ and $\bar{B}$ may be read from charts for various pad shapes presented previously(chapter 2 and reference[1]). For pads of any shape with thin constant land or sill width C, it is useful to note that:

$$\bar{A} \text{ thin land} = \frac{\text{total area} + \text{recess area}}{2 \times \text{total area}} \quad \ldots\ldots\ldots \quad (6)$$

$$\bar{B} \text{ thin land} = \frac{\text{total area} - \text{recess area}}{12C^2} \quad \ldots\ldots\ldots \quad (7)$$

Ideally, a pad is proportioned such that the sill width is 0·25 × the width. Under this condition, the approximate equations (6) and (7) over-estimate the load for a square pad by 7 per cent and the flow by 17 per cent. For thinner sill widths, the accuracy is better.

The values of $\bar{W}$ and $\bar{\lambda}$ given on the charts were computed from:

$$\bar{W} = \frac{P_1\,A_{e1} - P_2\,A_{e2}}{P_s\,(A_{e1} + A_{e2})} \quad \ldots\ldots\ldots \quad (8)$$

and

$$\bar{\lambda} = \frac{\bar{\lambda}_1\,A_{e1}}{A_{e1} + A_{e2}} + \frac{\bar{\lambda}_2\,A_{e2}}{A_{e1} + A_{e2}} \quad \ldots\ldots \quad (9)$$

The other relationships required for the computation were listed in Fig 2. 7 so that it should not be difficult for a designer to calculate more data. However it is not envisaged that this will prove necessary.

Explanatory notes

A hydrostatic bearing requires a control device between the source pressure and each pad. For most machine tool applications this device is simply a length of capillary tubing in the line to each pad. Capillary compensation has the advantage that the bearing clearance is independent of temperature variations as long as the temperature in the capillary is the same as in the pad. In other cases, the type of control device may be orifice, constant flow, or diaphragm valve, which all yield higher stiffness. A comprehensive description of control devices and their characteristics was given in chapter 3 and reference[1]

The method of design presented in this series allows the bearing to be specified without reference to dimensions of a control device. The information required to design the control device is known after completing the bearing procedure, which governs the order of making the calculations, that is, bearing first, control device second. Short detailed procedures for capillary and constant flow restrictors were given in the first chapter and therefore these are not repeated here.

The term 'design condition' is a useful notion which allows reference to the particular conditions when the bearing member is supported centrally in its clearance. The condition is denoted by a suffix *o* attached to relevant parameters, such that $p = p_o = \beta P_s$; $h = h_o, \bar{X}_o = 1$.

The data have all been computed for a value of the design pressure ratio $\beta = 0{\cdot}5$. It might be thought that there would be an advantage in selecting one value of β for the top pad and a different value for the bottom pad. This arrangement has the effect of displacing the bearing in the clearance for the design condition. However it does not appear to be an advantage since at one of the extremes of load the minimum gap is likely to be smaller than would otherwise be the case. A suitable design may be achieved simply by selecting the correct ratio of the top pad area to the lower pad area.

The minimum total area is calculated at Instruction 12. If the bearing is operating at sustained speeds sufficient to cause appreciable friction losses, then the choice of the actual area should be close to this value. In other cases it is permissible to employ a significantly larger area to obtain improved stiffness and load-carrying capacity.

Design procedure

Metric units are indicated in parentheses.

1. Write maximum positive load to be encountered [W positive] in units of lbf (N).
2. Write maximum negative load [W negative] in units of lbf (N).
3. Read the value of area ratio $E = A_1/A_2$ from Fig 2. (To be strictly correct, $E = A_{e1}/A_{e2}$, but because the upper and lower thrust pads would normally be of similar shape, the values A_1 and A_2 can be used).
4. Decide supply pressure P_s. Choose a low value for low temperature rise. Suggested value 300 lbf/in^2 (N/m^2).
5. Choose land width ratio for bearing pad. Recommended values see Fig 3 to 6.
6. Decide bearing clearance h_o (Total clearance is $2h_o$). Units in (m). Choose minimum value according to $h_o = 5 \times$ manufacturing tolerances on clearance.
7. Read flow shape factor for bearing pad $\bar{B}$ from Fig 4 or 6 as appropriate.
8. Read area shape factor for bearing pad $\bar{A}$ from Fig 3 or 5 as appropriate.
9. Read load parameter for positive loads [$\bar{W}$ positive] from Fig 7 to 9. $\bar{X}_2 = 1{\cdot}4$ for capillary and orifice control, $\bar{X}_2 = 1{\cdot}2$ for constant flow.
10. Read load parameter for negative loads [$\bar{W}$ negative] from Fig 7 to 9. $\bar{X}_2 = 0{\cdot}6$ for capillary and orifice control. $\bar{X}_2 = 0{\cdot}8$ for constant flow.
11. Calculate minimum total area requirement from

$$A \text{ minimum} = \frac{[W \text{ negative}]}{[\bar{W} \text{ negative}]\, P_s\, \bar{A}} \text{ in}^2 \text{ (m}^2\text{)}$$

12. Decide convenient value of total area A to be employed. In low speed applications, the actual area may

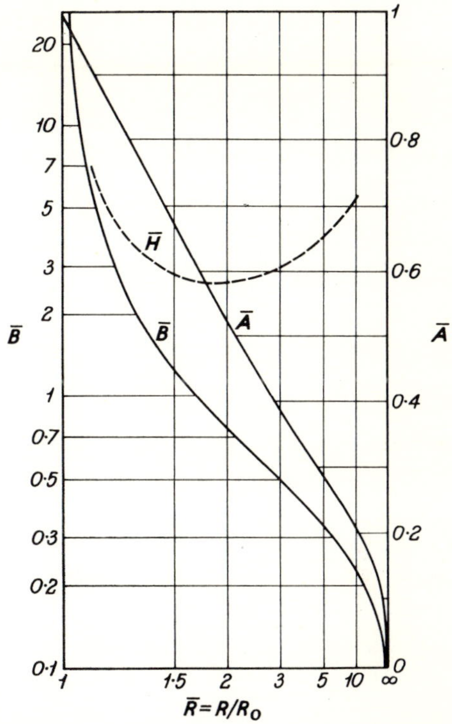

Fig 4

Pad coefficients for a circular thrust bearing. Recommended value R = 2

Fig 5

For a rectangular pad with uniform land width it is recommended that C/L = 0.25

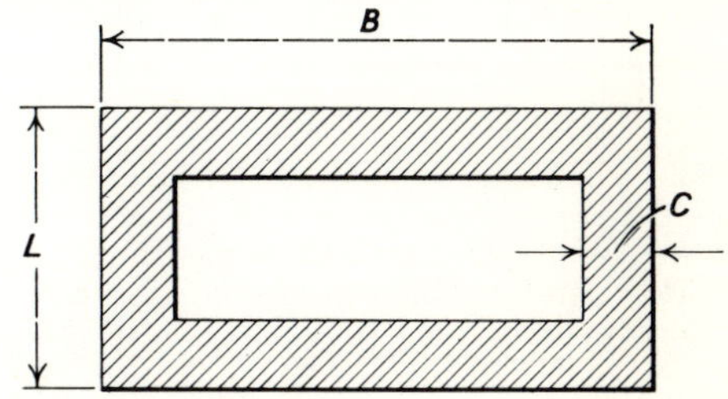

be considerably larger than A minimum if very high stiffness is required–in^2 (m^2).

13. Calculate the main pad area from

$$A_1 = \frac{A \, . \, E}{1 + E} \text{ in}^2 \text{ (m}^2\text{)}$$

14. Calculate other pad area from

$$A_2 = A - A_1 \text{ in}^2 \text{ (m}^2\text{)}$$

15. Calculate and list dimensions of the two pads–in (m).
16. Absolute viscosity η. A convenient value for many slow speed bearings is 5×10^{-6} reyns. Note viscosity

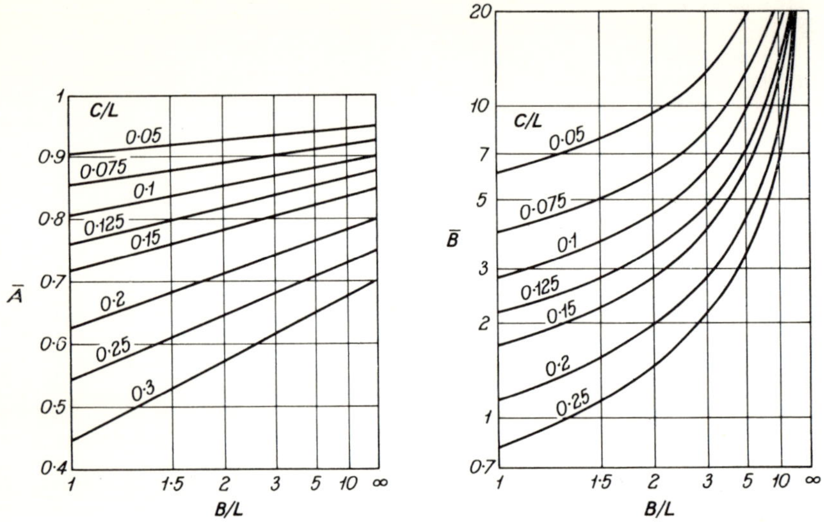

Fig 6

Coefficients for a rectangular thrust pad with equal land widths

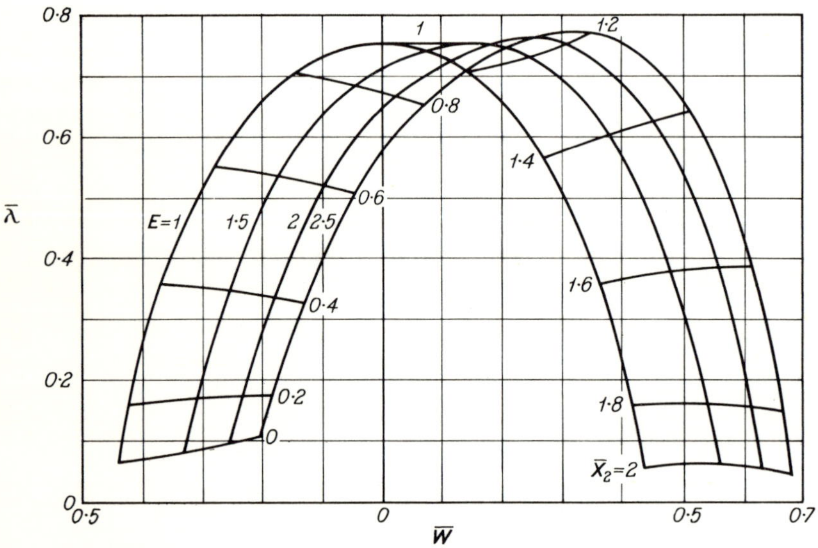

Fig 7

Capillary-controlled bearing characteristics

should be stated in a consistent set of units, i.e. reyns (lbf sec./in²) or N sec/m². To convert from units of cp to reyns multiply by $1{\cdot}45 \times 10^{-7}$

17. Calculate flow-rate q_o

$$q_o = \frac{2P_s h_o^3}{\eta}\ \beta\ \bar{B}\ \text{in}^3/\text{sec}\ (\text{m}^3/\text{sec})$$

18. Convert flow-rate to familiar units. Multiply flow in in³/sec $\times \frac{60}{277}$ to obtain gal/min.

19. Read stiffness parameter $\bar{\lambda}_o$ for design condition from chart, with $\bar{X} = 1$.

20. Calculate stiffness at design condition λ_o

$$\lambda_o = \frac{P_s\,(A_{e1} + A_{e2})}{h_o}\ .\ \bar{\lambda}_o$$

21. Read dimensionless gap $\bar{X}_1$ for

$$[\bar{W}\ \text{positive}] = \frac{[W\ \text{positive}]}{P_s A \bar{A}}$$

22. Calculate minimum value of h_1 from $h_1 = h_o \bar{X}_1$

23. Read dimensionless gap $\bar{X}_2$ for

$$[\bar{W}\ \text{negative}] = \frac{[W\ \text{negative}]}{P_s A A}$$

24. Calculate minimum value of h_2 from $h_2 = h_o\ \bar{X}_2$

25. Calculate maximum temperature difference ΔT between oil entering and oil leaving bearing from $\Delta T = 0{\cdot}0075\ P_s$ at $V = 0$ or very low speed (i.p.s. units) deg. F.

Example Square pad bearings to support 2,000 lbf. Under certain conditions load reverses to a maximum of 200 lbf. Other machine elements require a viscosity of 5×10^{-6} reyns. Capillaries to be used for compensation.

Fig 8

Orifice-controlled bearing characteristics

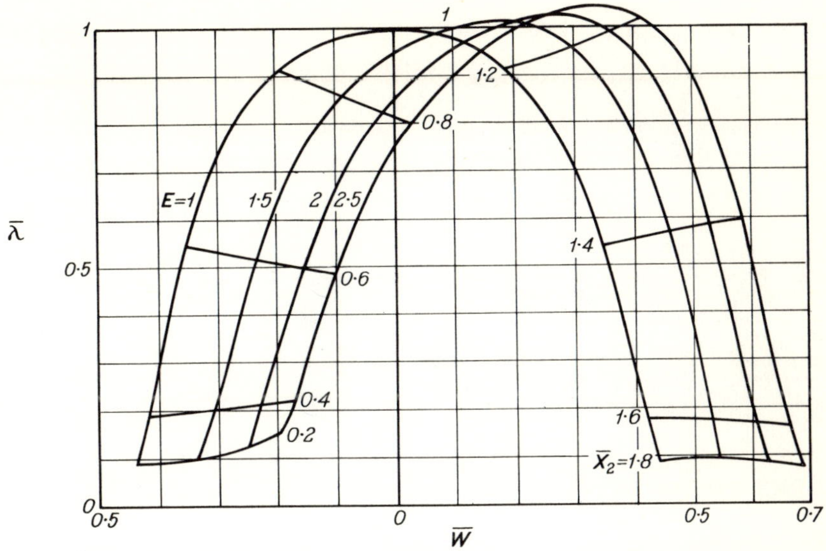

Fig 9

Constant-flow controlled bearing characteristics

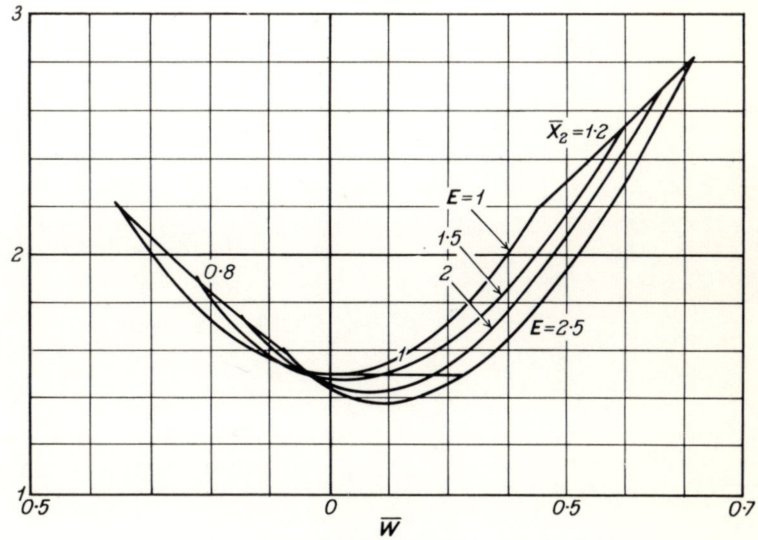

1. [W positive] $= 2{,}000$ lbf
2. [W negative] $= 200$ lbf
3. $E \quad = 2{\cdot}5$ approx.
4. $P_s \quad = 300$ lbf/in²
5. $C/L = 0{\cdot}25$
6. $h_o \quad = 0{\cdot}0014$ in
7. $\bar{B} \quad = 0{\cdot}72$
8. $\bar{A} \quad = 0{\cdot}54$
9. [$\bar{W}$ positive] $= 0{\cdot}5$
10. [$\bar{W}$ negative] $= 0{\cdot}05$
11. A minimum $= \dfrac{200}{0{\cdot}05 \times 300 \times 0{\cdot}54} = 24{\cdot}7$ in²
12. A actual $= 31{\cdot}5$ in² $=$ nearest convenient value
13. $A_1 = 31{\cdot}5 \times \dfrac{2{\cdot}5}{3{\cdot}5} = 22{\cdot}5$ in²
14. $A_2 = 31{\cdot}5 - 22{\cdot}5 = 9$ in²
15. $L_1 = \sqrt{2{\cdot}5} = 4{\cdot}75$ in
    $C_1 = 0{\cdot}25 \times 4{\cdot}75 = 1{\cdot}187$ in
    $L_2 = \sqrt{9} = 3$ in
    $C_2 = 0{\cdot}25 \times 3 = 0{\cdot}75$
16. $\eta \quad = 5 \times 10^{-6}$ reyns
17. $q_o \quad = \dfrac{2 \times 300 \times (0{\cdot}0014)^3}{5 \times 10^{-6}} \times 0{\cdot}5 \times 0{\cdot}72$
    $= 0{\cdot}12$ in³/sec
18. Flow-rate in familiar units $0{\cdot}12 \times \dfrac{60}{277}$
    $= 0{\cdot}025$ gal/min
19. $\bar{\lambda}_o \quad = 0{\cdot}75$ from Fig 7
20. $\lambda_o \quad = \dfrac{300 \times 31{\cdot}5}{0{\cdot}0014} \times 0{\cdot}75 = 5 \times 10^6$ lbf/in
21. [$\bar{W}$ positive] $= \dfrac{2{,}000}{300 \times 31{\cdot}5 \times 0{\cdot}54} = 0{\cdot}39$ for which

$\bar{X}_1 = 0{\cdot}76$ from Fig 7

22. Minimum value of $h_1 = 0{\cdot}0014 \times 0{\cdot}76 = 0{\cdot}00106$ in

23. $[\bar{W}\text{ negative}] = \dfrac{200}{300 \times 31{\cdot}5 \times 0{\cdot}54} = 0{\cdot}039$ for which

$\bar{X}_2 = 0{\cdot}6$

24. Minimum value of $h_2 = 0{\cdot}0014 \times 0{\cdot}6 = 0{\cdot}00084$ in

25. $\Delta T = 0{\cdot}0075 \times 300 = 2{\cdot}1/4$ deg F.

REFERENCES

1. J. P. O'Donoghue and W. B. Rowe: 'Hydrostatic Bearing Design', *Tribology*, Vol. 2, No. 1, February, 1969, pp. 25-71.
2. J P O'Donoghue and W B Rowe, 'Compensation Methods for Externally Pressurised Bearings', *Engineers Digest*, Vol. 30, No. 4, April 1969, pp. 49-61.
3. J P O'Donoghue and W B Rowe, 'Optimization of Hydrostatic Bearings for Minimum Power and Low Temperature Rise', *Industrial and Scientific Lubrication*, Vol. 21, April 1969, pp. 101-106.
4. J P O'Donoghue and W B Rowe, 'Hydrostatic Bearing Design–Supplement for Unequal Opposed Pads', *Tribology*, Vol. 2, No. 4, 1969, pp. 225-232.

Chapter Five

Combined journal-thrust bearings (the Yates bearing)

R S WEARING
J P O'DONOGHUE and W B ROWE

When a designer requires a bearing system to carry combined axial and radial loads there are four basic types of hydrostatic bearing arrangements to choose from:

1. A combination of separate thrust and journal bearings as shown at (*a*) in Fig. 1.
2. A conical bearing system as shown at (*b*).
3. A spherical bearing system as shown at (*c*).
4. A Yates type bearing as shown at (*d*).

The design methods for the first arrangement were given in the first two chapters. The thrust faces can be designed as an opposed combination using the design procedure detailed in chapter four. The conical system has been treated for the opposed case in chapter three, and the spherical bearing will be considered in the next chapter. Here, the design of the Yates type bearing is considered. As with the previous chapters, an attempt has been made to reduce the design problems to a minimum by offering a step-by-step procedure, together with the necessary design charts and tables.

The Yates type bearing was first described in a patent taken out in 1949, and the basic principle is to supply the thrust faces solely by the efflux from the journal bearing. A development of this basic principle is to incorporate a pocket in the thrust face, as shown in Fig. 2, a development which, as far as the authors can determine, was first published in a report of work carried out by Mechanical Technology, Inc, for the National Aeronautics and Space Administration, USA, in 1963. The advantages of this bearing were noted as being:

1. The pumping power per unit load is much less (50 per cent) than that for a separately-fed system.
2. The friction in the end thrust faces is much less than for a separately-fed thrust bearing.

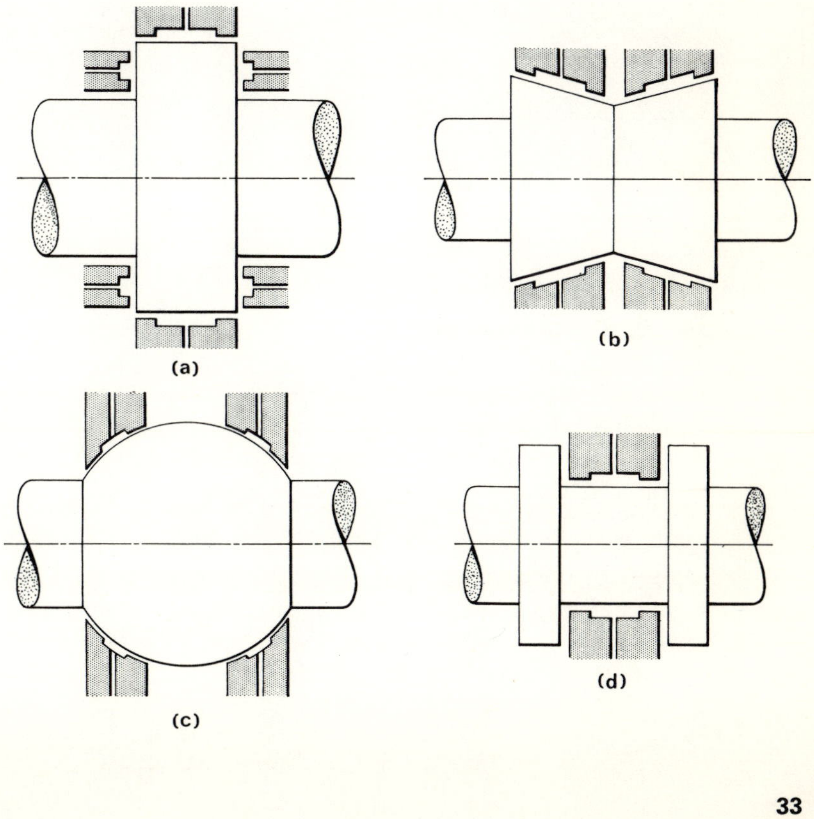

Fig 1

Diagrams representing the four basic hydrostatic bearing arrangements which will carry combined journal and axial thrust loads. Of these four, the Yates bearing shown at (d) has certain advantages and is the subject of this article

Symbols

P_s Supply pressure
P_r Recess pressure
o Suffix denoting design condition
h_o Axial clearance (single)
c Radial clearance
c_D Diametral clearance
D Shaft diameter
L Bearing length
D_i Inner land diameter of thrust bearing
D_o Outer land diameter of thrust bearing
$\bar{D}$ Ratio D_o/D_i
a Axial flow land width
b Circumferential flow land width
W Radial bearing load
T Thrust bearing load
$\beta_1 = P_1/P_s$ Pressure ratio of radial bearing at design condition
$\beta_2 = P_2/P_s$ Pressure ratio of thrust bearing at design condition
γ Circumferential flow factor for radial bearing
q Flow
B_{or} Flow factor for 1 pocket of radial bearing
B_{ot} Flow factor for the thrust bearing
$\bar{Q}$ Bearing flow factor
$\bar{W}$ Radial bearing load parameter
$\bar{T}$ Thrust bearing load parameter
A_{er} Effective area of radial bearing
A_{et} Effective area of thrust bearing (1 side only)
$\bar{A}_t$ Dimensionless area factor of thrust bearing for load (before central area subtracted)
H_f Friction power
H_p Pumping power

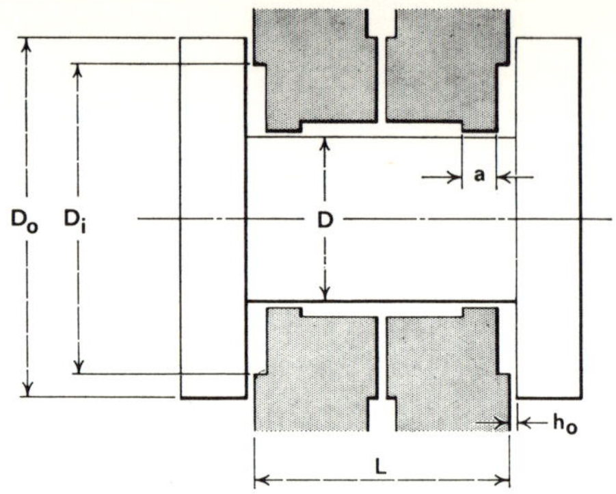

H_t Total power
η Viscosity
N Rotational speed (rev/sec)
$\bar{F}$ Friction power factor
A_{fr} Effective friction area of radial bearing
$\bar{A}_{f1}$ Dimensionless friction area of radial bearing
A_{ft} Friction area of thrust bearing
$\bar{A}_{f2}$ Dimensionless friction area of thrust bearing
n Number of pockets in radial bearing
S_H Optimization parameter $\frac{\eta N}{P_s}\left(\frac{D}{c_D}\right)^2$
λ Stiffness
$\bar{\lambda}$ Dimensionless stiffness
$\bar{\lambda}_j$ Journal stiffness parameter employed in the calculation of the dimensionless radial stiffness at the design condition ($\bar{\lambda}_{or}$).

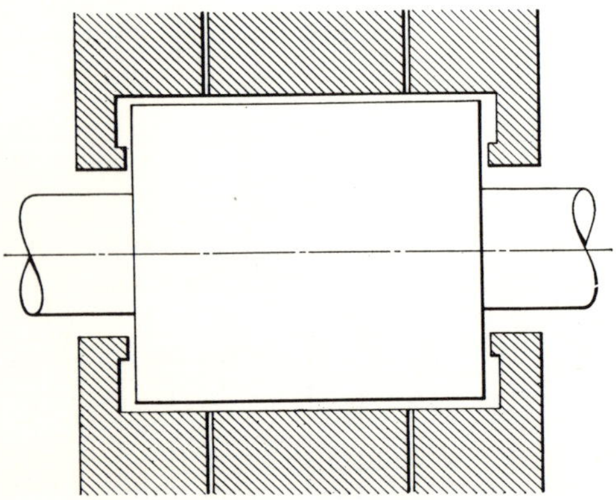

Fig 2

A development of the Yates bearing which incorporates a pocket in each thrust face

Fig 3 *A Yates bearing is here shown in the no-load condition (a), carrying axial thrust load (b), and carrying journal load (c). The variations in pressures are indicated by the lengths of the arrows*

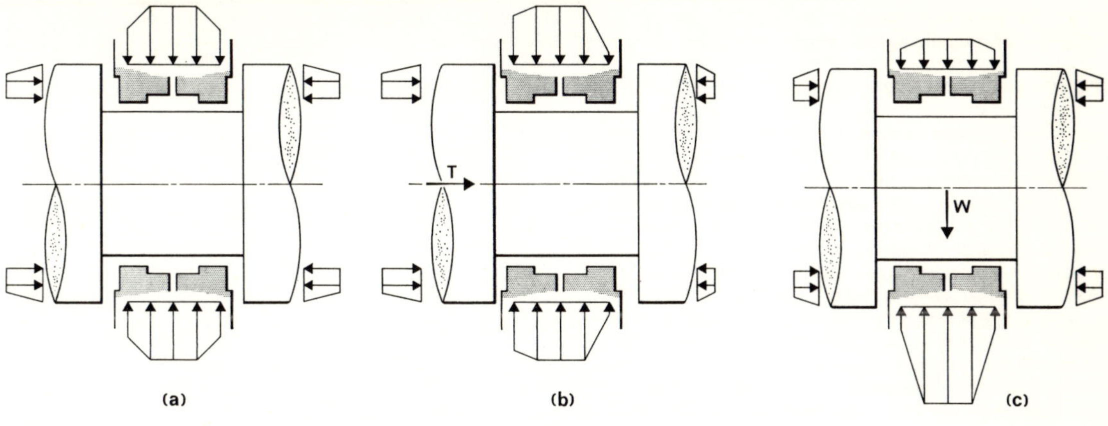

Fig 4

Curves at (a) show variations in pocket pressures under axial load, and at (b) under journal load

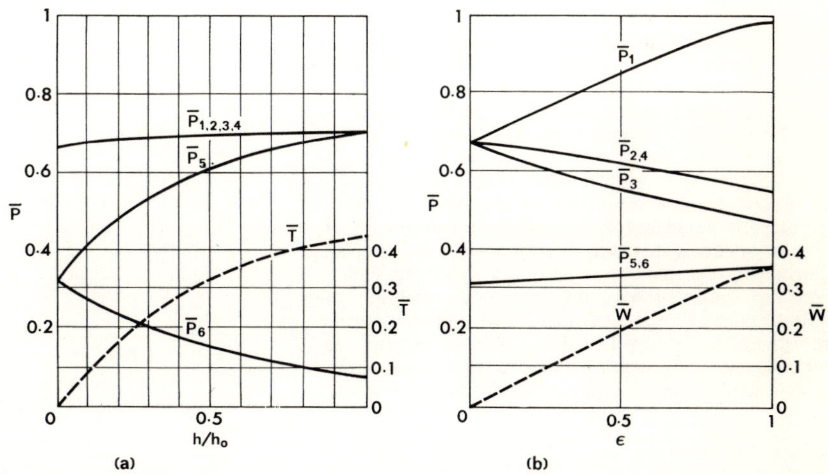

3. For gas bearings, the axial load capacity is greater than for a conventional thrust bearing. This statement may be true in the case of liquid-lubricated bearings, but will be achieved at the cost of a loss in axial dynamic stiffness and a reduction in the journal bearing performance.
4. The thrust bearings have no feed system, thus the bearing arrangement is simplified, the cost of machining the end faces is reduced, and the space requirements are less. Moreover, simplification of the system increases the reliability.

It should be noted from Fig 1 (a) and 2 that the thrust face can be either on an inward or outward flow arrangement from the journal. The floating cylinder arrangement described by the MTI workers has greater advantages for gas bearings than for liquid bearings, but if space is restricted, it is a useful option for the latter. A further application of this type of bearing for gear pumps has been found in the USA.

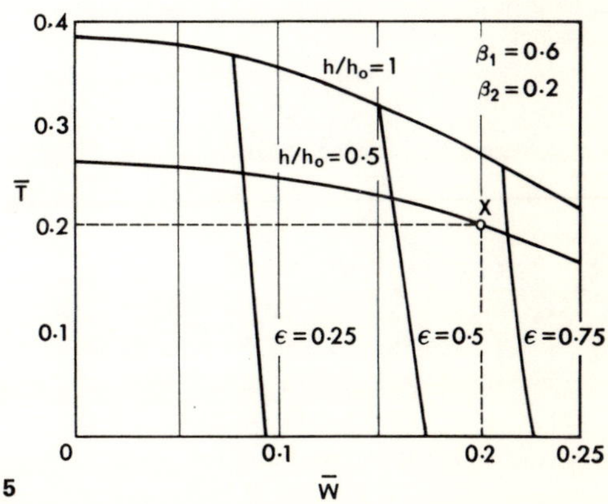

Fig 5

Curves showing the combined load characteristics for a Yates bearing

Since the term 'combined journal thrust' can be applied to both conical and spherical bearings, the authors propose to refer to the type of arrangement where the thrust faces are fed solely by the efflux from the journal bearing as a 'Yates bearing' after the original inventor of this most useful and efficient system.

Principle of operation

The principle of operation of a Yates bearing is illustrated in Fig 3, and pressure distribution under no-load conditions is shown at (*a*). Journal pockets are fed through a suitable control arrangement, and the capillary control has special advantages for this application (over those given in ref 3) because the journal forms part of the control to the thrust and is by its nature a viscous flow restrictor. The fluid flows axially across the lands into the thrust pockets, and then radially across the thrust lands to exhaust at atmospheric pressure. When an axial load is applied, as illustrated at (*b*) in Fig. 3, the increased flow resistance on the loaded side causes the pressure to rise, while the lessened resistance at the opposite face causes the pressure to be reduced. The difference in the thrust face pressures gives the axial load capacity. When a radial load is applied, the pressures in the journal pockets vary, but the side pocket pressures are virtually constant, as shown at (*c*) in Fig. 3.

These variations in pocket pressures are shown for a particular bearing in Fig 4. The axial load variation is indicated by the curves at (*a*) and the radial load variation by the curves at (*b*). Combined load characteristics for this typical bearing are shown in Fig 5. As can be seen from this graph, there is a reduction in the radial load capacity with applied thrust, and similarly the thrust load capacity is reduced by the presence of a radial load. The authors have computed a wide range of solutions for this bearing arrangement, and have developed the following design procedure which is based on the worst condition of simultaneous maximum values of radial and axial loads and is therefore conservative. The effects of rotation are beneficial, and at optimum conditions, they result in an increased load capacity of approximately 20 per cent.

Basic relationships

The notation has been kept as consistent as possible with that used in the previous chapters. The load parameter is defined by:

$$\bar{W} = \frac{W}{LDP_s} \quad \ldots\ldots\ldots\ldots\ldots \quad (1)$$

The flow parameter is defined by,

$$\bar{Q} = \frac{q\,\eta}{P_s h_o^3} \quad \ldots\ldots\ldots\ldots\ldots \quad (2)$$

The stiffness parameters are defined by,

$$\bar{\lambda}_{or} = \frac{\lambda_{or}\, c_D}{2P_s\, A_{er}} \quad \ldots\ldots\ldots\ldots \quad (3a)$$

$$\bar{\lambda}_{ot} = \frac{\lambda_{ot}\, h_o}{P_s\, A_{et}} \quad \ldots\ldots\ldots\ldots \quad (3b)$$

The friction power factor is defined by,

$$\bar{F} = \frac{H_f\, c_D}{\eta\, N^2\, D^4} \quad \ldots\ldots\ldots\ldots \quad (4a)$$

and $\bar{F}$ can be calculated from

$$\bar{F} = \pi^3\,(2\,.\,\frac{L}{D}\,.\,\bar{A}_{f1} + 2\,\frac{c_D}{h_o}\,.\,\bar{A}_{f2}) \quad \ldots\ldots \quad (4b)$$

the two friction areas being related to the bearing dimensions by

$$\bar{A}_{f1} = \frac{A_{fr}}{\pi\, D\,.\,L} \quad \ldots\ldots\ldots\ldots \quad (5a)$$

$$\bar{A}_{f2} = \frac{A_{ft}}{\pi D^2} \quad \ldots\ldots\ldots\ldots\ldots \quad (5b)$$

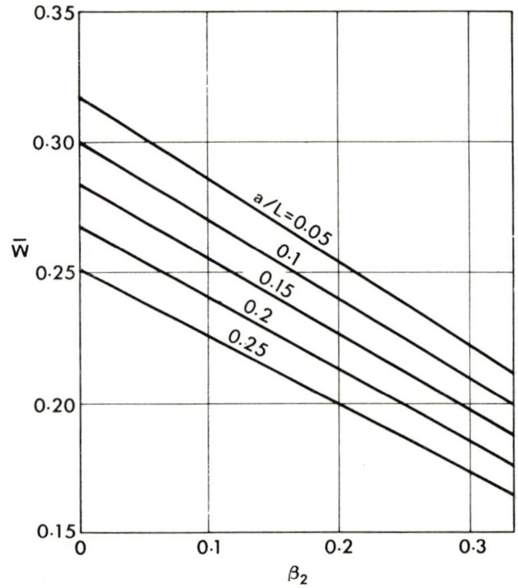

Fig 6

The value of the radial load parameter $\bar{W}$ is read from this nomogram

Design procedure

1. Specify the maximum load (radial), W (lbf)
2. Specify the maximum axial load, T (lbf)
3. Specify the maximum operating speed, N (rev/s)
4. Specify the supply pressure, P_s (lbf/in²)

Notes: The factors affecting the selection of supply pressure are,

(i) The cost of the pumping system increases with pressure. A relatively inexpensive gear pump can be used for pressures up to 400 lbf/in² (approx).

(ii) The temperature rise of an optimized bearing is directly proportional to supply pressure. Since viscosity varies with both pressure and temperature, and viscosity variations decrease the capacity for an

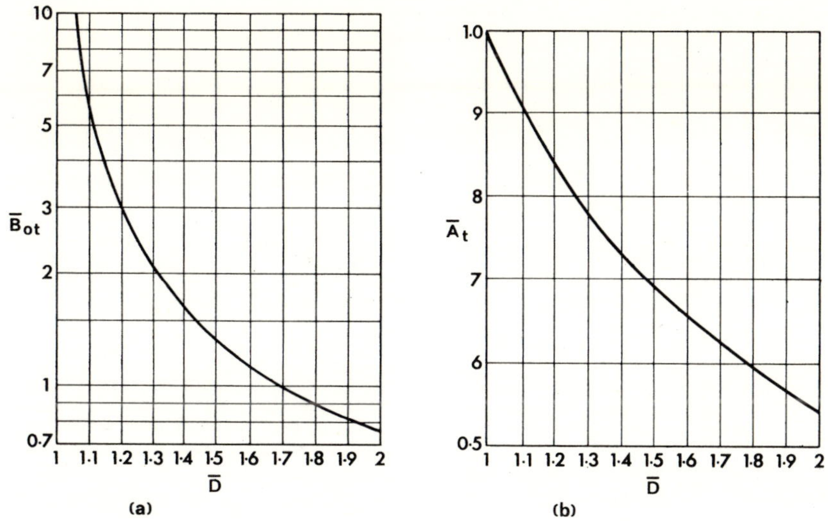

Fig 7

Value of the thrust land ratio (D_0/D_1) is read from the chart (a) and the thrust load parameter ($\bar{A}_t$) from chart (b)

optimized bearing, high pressures should be avoided if possible.

(iii) If the bearing size is limited, then the pressure will be determined by the load values.

5. The value of the thrust pocket pressure ratio, β_2, must now be chosen. For maximum axial stiffness, the value of β_2 should be 0·5, but the radial load capacity will be reduced by 50 per cent. The radial load capacity and stiffness will be virtually unaffected by use of a low value of β_2, say 0·05, but although the thrust load capacity will be satisfactory, the concentric stiffness will be low. Where both axial and radial stiffness are required, a compromise value for β_2 of 0·33 should be chosen. This procedure is limited therefore over the practical range of β_2 values between 0 and 0·33. An advantage of using a low value of β_2 is that the thrust face friction will be low.

6. The value of the journal pocket pressure ratio is calculated,

$$\beta_1 = \frac{1 + \beta_2}{2} \quad \ldots\ldots\ldots\ldots \quad (6)$$

7. The axial land ratio, a/L, is now specified. As in the case of conventional journal bearings, this ratio should initially be taken as 0·25. At high speed, this ratio may result in the value of viscosity calculated at step 33 being too low, in which case the value may be reduced.

8. The value of the radial load parameter, $\bar{W}$ is read from Fig 6.

9. The value of shaft diameter is determined either from other design requirements (such as torsional or bending stiffness) or it may be calculated from,

$$D = \sqrt{\frac{W}{P_s \bar{W}}} \text{ (in)} \quad \ldots\ldots\ldots\ldots \quad (7)$$

10. The minimum value of the shaft length is calculated from,

$$L = \frac{W}{P_s D \,.\, \bar{W}} \text{ (in)} \quad \ldots\ldots\ldots\ldots \quad (8)$$

If the value thus calculated is such that the length to diameter ratio, L/D, is less than 0·25, then it should be increased to this value, and the values of supply pressure P_s and diameter D should be reviewed. If the value for the length is such that the L/D ratio is greater than 1·5, then the pressure or diameter must be increased to reduce the length. Ideally, the length-to-diameter ratio should be approximately unity, to give a good compromise between the needs of low flow and high load capacity, and the difficulties arising from the axial alignment tolerances reducing the permitted eccentricity of the shaft.

11. The number of pockets may now be determined. Where the L/D ratio is less than 0·75, the minimum number of pockets is six. Above this value, the minimum number of pockets is four. The advantages of a larger number of pockets are: increased load capacity, increased stiffness, and less variation of characteristics with direction of applied load. The disadvantage is the increased machining costs and additional complexity of the supply system.

12. The value of the interpocket land width, b, is calculated,

$$b = \frac{4\pi D}{3n} \cdot \frac{a}{L} \text{ (in)} \quad \ldots\ldots\ldots \quad (9)$$

where n is the number of pockets.

13. The pocket flow coefficient, $\bar{B}_{or}$, is calculated,

$$\bar{B}_{or} = \frac{\pi D}{6\,na} \quad \ldots\ldots\ldots\ldots \quad (10)$$

14. The value of the radial clearance, c, is now decided, (in).

15. The value of the thrust face clearance, h_o, is now specified, (in).

16. The axial flow coefficient is now calculated from,

$$\bar{B}_{ot} = \frac{n\bar{B}_{or}}{2}\left(\frac{\beta_1}{\beta_2} - 1\right) \cdot \left(\frac{c}{h_o}\right)^3 \quad \ldots\ldots \quad (11)$$

17. The value of the thrust land ratio, $\bar{D}$, which is the ratio of the outer thrust face diameter, D_o, over the inner thrust land diameter, D_i, is read from chart (*a*) Fig 7, knowing the value of $\bar{B}_{ot}$. For values of $\bar{B}_{ot}$ greater than 10, the value of $\bar{D}$ may be calculated from,

$$\bar{D} = 1 - \frac{2\pi}{12 \,.\, B_{ot} - \pi} \quad \ldots\ldots\ldots\ldots \quad (12)$$

18. The value of $\bar{T}$, the thrust parameter, is read from Fig 8, against the chosen value of β_2.

19. The required effective thrust area, A_{et}. is now calculated from,

$$A_{et} = \frac{T}{P_s \,.\, \bar{T}} (\text{in}) \quad \ldots\ldots\ldots\ldots\ldots\ldots \quad (13)$$

20. The value of the thrust load parameter, $\bar{A}_t$, is read from chart (*b*) Fig 7. This value is not the actual pad coefficient, it is the coefficient of a pad with an identical land diametral ratio but with no loss of central area as with the annular pad of the Yates bearing.

21. The outside diameter of the thrust face, D_o, can now be calculated,

$$D_o = \sqrt{\frac{4 \,.\, A_{et}}{\pi \,.\, \bar{A}_t} + \frac{D^2}{\bar{A}_t}} (\text{in}) \quad \ldots\ldots\ldots\ldots \quad (14)$$

22. The internal land diameter, D_i, is calculated,

$$D_i = D_o/\bar{D} (\text{in}) \quad \ldots\ldots\ldots\ldots\ldots \quad (15)$$

Note: This value must not be less than the shaft diameter, D. If it is less, then it is made equal to D, and D_o is made equal to $D \times \bar{D}$. Alternatively, β_2 may be reduced.

23. The value of the circumferential flow factor, γ, is given by,

$$\gamma = \frac{n \,.\, a(L - a)}{\pi D \,.\, b} \quad \ldots\ldots\ldots\ldots \quad (16)$$

Fig 8

The value of $\bar{T}$, the thrust parameter, is obtained from this chart

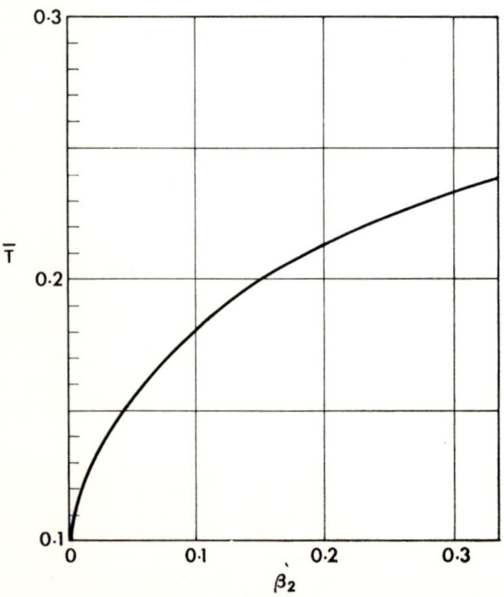

24. The value of the journal stiffness parameter is determined by using the relationships given in the accompanying table. This factor neglects the presence of the thrust pad and is termed $\bar{\lambda}_j$.

Dimensionless stiffness coefficient, $\bar{\lambda}_j$, to be used in step 24

No of pockets	Capillary	Orifice
3	$\frac{0 \cdot 27}{1 - 0 \cdot 75\gamma}$	$\frac{0 \cdot 54}{1 \cdot 5 - 1 \cdot 5\gamma}$
4	$\frac{0 \cdot 955}{1 - 0 \cdot 5\gamma}$	$\frac{1 \cdot 91}{1 \cdot 5 - \gamma}$
5	$\frac{1 \cdot 03}{1 - 0 \cdot 345\gamma}$	$\frac{2 \cdot 125}{1 \cdot 5 - 0 \cdot 69\gamma}$
6	$\frac{1 \cdot 075}{1 - 0 \cdot 25\gamma}$	$\frac{2 \cdot 15}{1 \cdot 5 - 0 \cdot 5\gamma}$

25. The actual stiffness parameter, $\bar{\lambda}_{or}$, is calculated,

$$\bar{\lambda}_{or} = \bar{\lambda}_t (1 - \beta_2) \quad \ldots\ldots\ldots\ldots \quad (17)$$

26. The journal stiffness at the concentric position is calculated by rearrangement of equation 3a,

$$\lambda_{or} = \frac{2 P_s \,.\, A_{er}}{c_D} \bar{\lambda}_{or} (\text{lbf/in}) \quad \ldots\ldots\ldots \quad (18)$$

where c_D is the diametral clearance (ie, $2 \,.\, c$) and A_{er} is equal to $D(L - a)$

27. The value of $\bar{\lambda}_{ot}$, the axial stiffness parameter, is read from Fig 9.

28. The concentric axial stiffness is calculated by rearrangement of equation 3b,

$$\lambda_{ot} = \frac{P_s \,.\, A_{et}}{h_o} \bar{\lambda}_{ot} (\text{lbf/in}) \quad \ldots\ldots\ldots\ldots \quad (19)$$

29. The thrust friction area coefficients are calculated from,

$$\bar{A}_{f2} = \frac{D_o^4 - D_i^4}{8D^4} + \frac{D_i^4 - D^4}{160D^4} \quad \ldots\ldots \quad (20)$$

or they may be read from Fig 10.
The journal friction coefficient $\bar{A}_{f1}$ is read from Fig 11.

30. The thrust friction power parameter is calculated from,

$$\bar{F} = \pi^3 \left\{ 2 \,.\, \frac{L}{D} \,.\, \bar{A}_{f1} + 2 \,.\, \frac{c_D \bar{A}_{f2}}{h_o} \right\} \quad \ldots\ldots \quad (21)$$

31. The flow coefficient, $\bar{Q}$, is calculated from,

$$\bar{Q} = 2 \,.\, \beta_2 \,.\, \bar{B}_{ot} \quad \ldots\ldots\ldots \quad (22)$$

32. The optimization parameter, S_H, is calculated,

$$S_H = \sqrt{\frac{\bar{Q}}{8 \,.\, \bar{F}} \left(\frac{h_o}{c} \right)^3} \quad \ldots\ldots\ldots \quad (23)$$

33. The optimum viscosity is calculated from,

$$\eta = \frac{S_H \,.\, P_s}{N} \left(\frac{c_D}{D} \right)^2 (\text{reyn}) \quad \ldots\ldots\ldots\ldots \quad (24)$$

Notes: If the value of the viscosity, η, thus calculated is too low, then the first step is to reduce the land-width, a, and then if this is less than $100 \times c$, the clearance must be increased. The calculations are then repeated from step 7. If the value of viscosity is too high, then it is set at a practicable maximum value and the procedure is continued using this value.

34. The flow is calculated from,

$$q = \frac{P_s \cdot h_o^3}{\eta} \bar{Q} \text{ (in}^3\text{/s)} \quad \dots\dots\dots\dots \quad (25)$$

35. The flow is converted to familiar units,

$$q' = q \cdot \frac{60}{277} \text{ (gal/min)} \quad \dots\dots\dots\dots \quad (26)$$

36. The pumping power loss is given by,

$$H_p = \frac{P_s \cdot q}{6{,}600} \text{ (hp)} \quad \dots\dots\dots\dots \quad (27)$$

37. The friction power loss is given by,

$$H_f = \frac{\eta N^2 D^4}{6{,}600 c_D} \bar{F} \text{ (hp)} \quad \dots\dots\dots\dots \quad (28)$$

38. The total power loss is given by,

$$H_t = H_p + H_f \text{ (hp)} \quad \dots\dots\dots\dots \quad (29)$$

39. Now calculate a suitable restrictor using the procedure given in chapter one for a flow equal to the total flow divided by the number of pockets in the journal bearing, ie q/n.

Example:

The same values of thrust and radial load as for the conical example are used to illustrate this method.

1. $W = 500$ lbf
2. $T = 300$ lbf
3. $N = 20$ rev/s (ie 1,200 rev/min)
4. $P_s = 300$ lbf/in^2
5. $\beta_2 = 0{\cdot}2$
6. $\beta_1 = 0{\cdot}6$
7. $\frac{a}{L} = 0{\cdot}25$
8. $\bar{W} = 0{\cdot}2$
9. $D = \sqrt{500/(300 \times 0{\cdot}2)} = 2{\cdot}875$ in
10. $L = 2{\cdot}875$ in
11. $n = 4$
12. $b = (4 \times \pi \times 2{\cdot}875 \times 0{\cdot}25)/(3 \times 4) = 0{\cdot}75$ in
13. $\bar{B}_{or} = (\pi \times 2{\cdot}875)/(6 \times 4 \times 0{\cdot}72) = 0{\cdot}522$
14. $c = 0{\cdot}001$ in, $c_D = 0{\cdot}002$ in
15. $h_o = 0{\cdot}001$ in
16. $\bar{B}_{ot} = \frac{4 \times 0{\cdot}522}{2} \times \left(\frac{0{\cdot}6}{0{\cdot}2} - 1\right) \times \left(\frac{0{\cdot}001}{0{\cdot}001}\right)^3 = 2{\cdot}09$
17. $\bar{D} = 1{\cdot}3$
18. $\bar{T} = 0{\cdot}21$
19. $A_{et} = 300/(300 \times 0{\cdot}21) = 4{\cdot}77$ in^2 (for one thrust face only)
20. $\bar{A}_t = 0{\cdot}8$
21. $D_o = (4 \times 4{\cdot}77)/(\pi \times 0{\cdot}8) + 2{\cdot}875^2/0{\cdot}8 = 4{\cdot}15$ in
22. $D_i = 4{\cdot}15/1{\cdot}3 = 3{\cdot}25$ in
23. $\gamma = [4 \times 0{\cdot}72 \times (2{\cdot}875 - 0{\cdot}72)]/(\pi \times 2{\cdot}875 + 0{\cdot}75) = 0{\cdot}914$
24. $\bar{\lambda}_j = 0{\cdot}955/(1 - 0{\cdot}914/2) = 0{\cdot}656$
25. $\bar{\lambda}_{or} = 0{\cdot}656 \times (1 - 0{\cdot}2) = 0{\cdot}525$
26. $\lambda_{or} = (2 \times 300 \times 2{\cdot}875 \times (2{\cdot}875 - 0{\cdot}72) \times 0{\cdot}525)/0{\cdot}002 = 9{\cdot}75 \times 10^5$ lbf/in
27. $\bar{\lambda}_{ot} = 0{\cdot}8$ (note: this value is based on the effective area of one pad)
28. $\lambda_{ot} = (300 \times 4{\cdot}77 \times 0{\cdot}8)/0{\cdot}001 = 1{\cdot}15 \times 10^6$ lbf/in
29. $A_{f2} = (4{\cdot}15^4 - 3{\cdot}25^4)/8 \times 2{\cdot}875^4 - (3{\cdot}25^4 - 2{\cdot}875^4)/160 \times 2{\cdot}875^4 = 0{\cdot}34 - 0{\cdot}004 = 0{\cdot}344$, $A_{f1} = 0{\cdot}7$ (from Fig 11).
30. $\bar{F} = \pi^3 \times (2 \times 1 \times 0{\cdot}7 + 2 \times \frac{0{\cdot}002}{0{\cdot}001} \times 0{\cdot}344) = 86$
31. $\bar{Q} = 2 \times 0{\cdot}2 \times 2{\cdot}09 = 0{\cdot}836$
32. $S_H = \sqrt{(0{\cdot}836 \times 0{\cdot}001^3)/(8 \times 86 \times 0{\cdot}001^3)} = 0{\cdot}035$
33. $\eta = (0{\cdot}035 \times 300 \times 0{\cdot}002^2)/(20 \times 2{\cdot}875^2) = 2{\cdot}52 \times 10^{-7}$ reyn ie $1{\cdot}75$ cp
34. $q = 300 \times 0{\cdot}001^3 \times 0{\cdot}836/2{\cdot}52 \times 10^{-7} = 1$ in^3/s
35. $q' = 1 \times 60/277 = 0{\cdot}216$ gal/min
36. $H_p = 300 \times 1/6600 = 0{\cdot}0455$ hp
37. $H_f = 2{\cdot}52 \times 10^{-7} \times 20^2 \times 2{\cdot}875^4 \times 86/(0{\cdot}002 \times 6600) = 0{\cdot}046$ hp

    (*Note:* for an optimized bearing these values should be equal, and this condition provides a useful check in the procedure.)
38. $H_t = 0{\cdot}0455 - 0{\cdot}046 = 0{\cdot}0915$ hp

If a restriction is imposed, as in the previous conical case (3), that the minimum viscosity is $1{\cdot}2 \times 10^{-6}$ reyn, then the land ratio would be reduced to 0·05 and the calculation repeated from step 7.

The results show that for the Yates type bearing the stiffnesses are increased, the power is of the same order, and the number of supply feeds to the system is halved, compared with a twin-feed bearing arrangement. The Yates type bearing is useful in applications where it is not possible to have two conical bearings with an adequate separation distance between them. Furthermore, the conical bearing and the Yates bearing require considerably less power than the separate thrust and journal bearing arrangements.

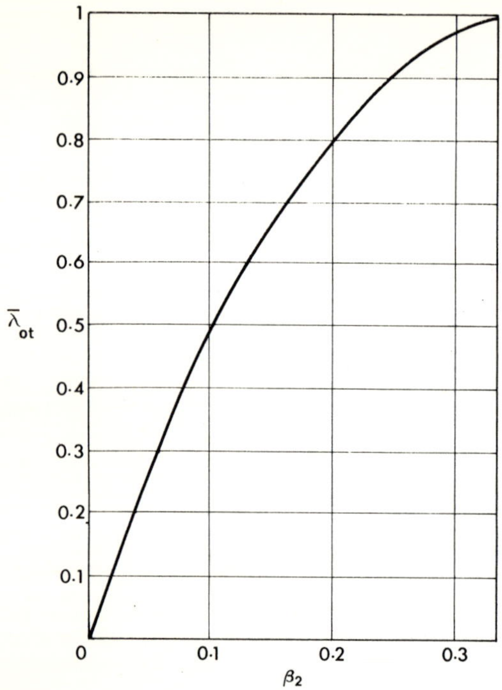

Fig 9

Value of the axial stiffness parameter ($\bar{\lambda}_{ot}$) is read from this chart

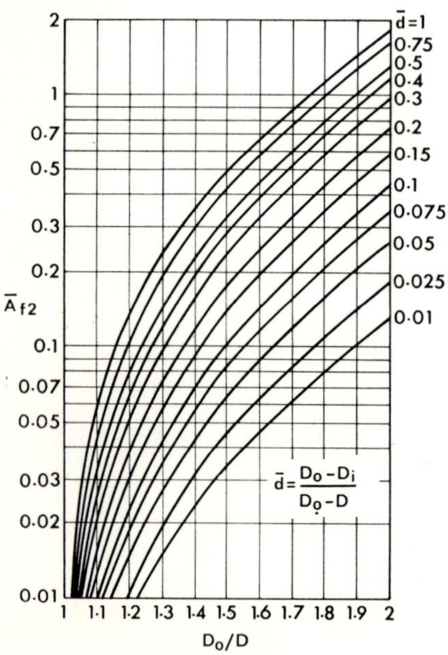

Fig 10

Curve from which friction area coefficient ($\bar{A}_{f2}$) can be read

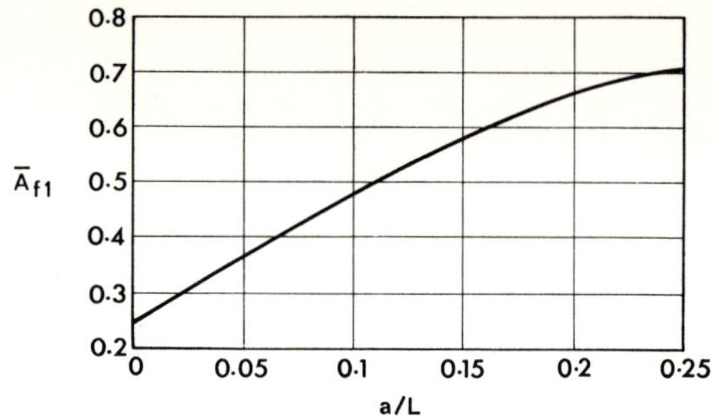

Fig 11

The value of the journal friction coefficient $\bar{A}_{f1}$ is read from this curve

REFERENCES

1 H G Yates, Combined Journal and Thrust Bearings. *UK Patent No 639,293, June 50*

2 J W Lund, Static Stiffness and Dynamic Angular Stiffness of the Combined Hydrostatic Journal-Thrust Bearing. *Mechanical Technology Inc Rept No MTI-63TR45, 1963.* (A less detailed description is contained in NASA SP 5058, available from the NASA Office of Technological Utilization).

3 J P O'Donoghue and W B Rowe, Hydrostatic Bearing Design. *Tribology, Vol. 2, pp. 25-71, February, 1969.*

4 J P O'Donoghue and W B Rowe, Compensation Methods for Externally Pressurised Bearings, *Engineers Digest, Vol. 30, pp. 49-61, April 1969.*

Chapter Six

Multi-recess bearings

J P O'DONOGHUE and W B ROWE

In chapter two of this book the authors presented a design method for plane hydrostatic bearings. A disadvantage of the single recess hydrostatic bearing is that when used in isolation it offers virtually no resistance to tilting moments, and a very small offset of the applied load can cause the bearing to tilt and can result in metal to metal contact. The designer can overcome this disadvantage by using either a multiple bearing system as shown in Fig 1, or, if there is a restriction on available space, a multi-recess bearing as shown in Fig 2. The multi-recess bearing has the advantage that the inter-pocket lands have little effect on the load capacity of the bearing and the available area is thus virtually identical to that of the single pocket bearing. Multiple bearings reduce the effective bearing area, but have a greater resistance to tilt. Methods of design of multiple bearing arrangements have been given in a more extensive article.[1]

Here, as in the previous chapters, the aim has been to simplify the calculations as far as possible and to present the necessary computed information in simple chart form.

It should be clearly understood that each recess of a multi-recess bearing must be independently controlled by a suitable compensating mechanism which is essential to all hydrostatic bearings. In this procedure it is assumed that oil may be fed to each recess either through a separate capillary tube or through a separate orifice, and it is recommended that the dimensions are such that the recess pressure is equal to one half the supply pressure under design conditions (i.e. $\beta = 0{\cdot}5$).

It is also possible to control these bearings with other systems, for example, by providing a separate constant flow valve or pump for each recess, or by the use of diaphragm valves. These systems are listed in order of ascending stiffness. A bearing designed conservatively for capillary control may be used satisfactorily with these other arrangements, but orifices and constant flow compensation introduce temperature dependence. The significance of temperature dependence is that if the system temperature changes there is a consequent variation in viscosity, and hence in flow-rate, so that the gap size changes even though the load remains constant.

The order of design is that the designer specifies the bearing pad first and then carries out the procedure for the control arrangement.

In the following procedure, the designer specifies the load that the bearing must support and a suitable clearance. The basis for choosing clearance is that large clearance results in excessive flow and low stiffness, whereas too small clearance requires close tolerances and allows very little margin for deviation from the specified condition. It is therefore suggested that clearance should be 5 to 10 times the manufacturing tolerance on clearance. Values for supply pressure, tilting, stiffness and flow-rate are calculated. The maximum tilting stiffness is estimated conservatively.

It is not necessary to follow the procedure rigidly. The designer may prefer initially to follow the basic procedure outlined in chapter two for plane bearings, which is also applicable to the multi-recess arrangement. However, the latter configuration is chosen because of the ability to withstand tilting moments, and therefore the plane bearing procedure should be followed by the estimation of tilts and moments from this chapter. The validity of these charts is unaffected provided the designer ensures that $\bar{W}$ remains within the range 0·35 to 0·65 for capillary compensation, or within the range 0·45 to 0·75 for orifice compensation.

The procedure for tilting pads also involves calculation of the critical velocity up to which speed friction and hydrodynamic effects will not be excessive. Critical velocity may be raised by reducing viscosity, reducing land widths or increasing clearance, which is the correct order for making adjustments to the design.

Basic relationships

The load relationship is given by

$$W_o = P_s A \bar{A} \beta \qquad (1)$$

The flow rate is given by

$$q_o = \frac{P_s h_o^3}{\eta} \beta \bar{B}_o \qquad (2)$$

The stiffness is given by

$$\lambda = \frac{P_s A}{h_o} \bar{A} \bar{\lambda} \qquad (3)$$

The tilting stiffness is given by

$$k_o = \frac{P_s A L^2}{2h_o} \bar{K}_o \qquad (4)$$

The values for $\bar{A}$, $\bar{B}$, and $\bar{K}$ are shown in Fig 6.3 - 6.6 and the values for $\bar{\lambda}$ are as given in chapter two. The values for $\bar{K}$ are given for a pressure ratio of 0·5, this being a useful compromise value between load and stiffness as explained in reference[1]. At the design condition the values of the stiffness parameter are then

$$\left.\begin{array}{l}\bar{\lambda}_o = 0{\cdot}75 \text{ for capillary control} \\ \bar{\lambda}_o = 1{\cdot}00 \text{ for orifice control}\end{array}\right\} \qquad (5)$$

Design procedure

The design procedure is best illustrated by a specific example. In this case study a bearing is required to support a load of 1500 lbf. The bearing size is limited to 1·5 by 4 in, the clearance is chosen to be 0·0015 in, and the lubricant has a viscosity of 25×10^{-6} reyn.

(1) The land-width ratio is initially chosen as 0·125.

Symbols

P_r Recess pressure
P_s Supply pressure
W Supported load on bearing thrust
A Total projected bearing area
A_e Effective area over which P_r may be assumed to act
$\bar{A} = A_e/A$ Area shape factor (or load factor) for bearing pad
h Bearing gap
η Absolute or dynamic viscosity
$\bar{B} = \frac{q\eta}{P_r h^3}$ Flow shape factor for bearing pad
q Flow rate
o Suffix denoting designed operating conditions
1 Suffix denoting tilt about axis 1 (Fig 2)
2 Suffix denoting tilt about axis 2 (Fig 2)
$\bar{W} = \frac{W}{P_s A_e}$ Load parameter
$\lambda =$ Stiffness against direct thrust
$\bar{\lambda} = \frac{\lambda h_o}{P_s A_e}$ Dimensionless stiffness
$k =$ Tilting stiffness per radian deflection
$\bar{K} = \frac{2k_o h_o}{P_s A L^2}$ Tilting stiffness parameter
$\beta = \frac{P_{ro}}{P_s}$ Pressure ratio for designed operating condition without tilt
L Longer outside dimension of rectangular multi-recess pad
B Shorter outside dimension of rectangular multi-recess pad
C Land width
A_f Effective friction area
H_p Pumping power
H_f Friction power
V_c Critical velocity below which hydrodynamic effects will not be dangerous
ρ Mass density of oil
S.G. Specific gravity of oil
R_e Reynolds number
C_f Orifice coefficient
K_c Capillary constant
K_{or} Orifice constant

Fig 1 below

Multiple bearing system

Fig 2 right

Multi-recess bearing system

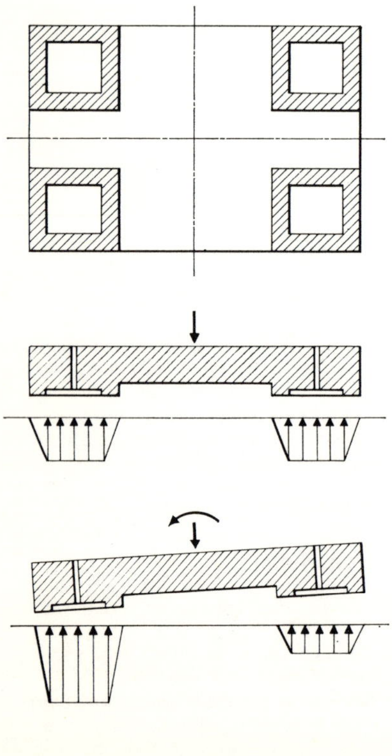

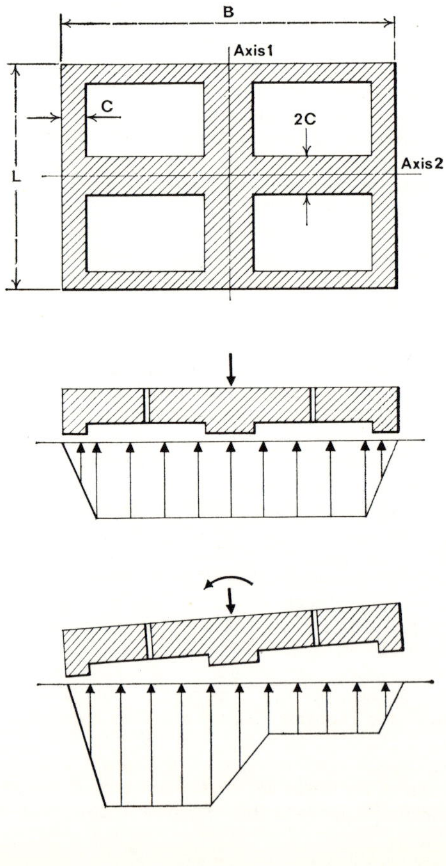

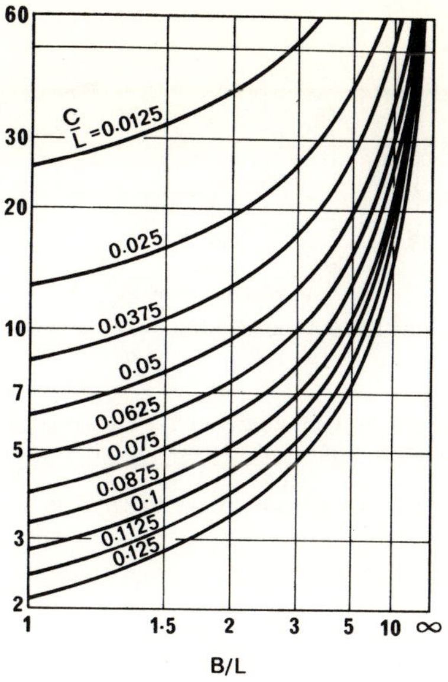

Fig 3

Load parameter for rectangular multi-recess pads

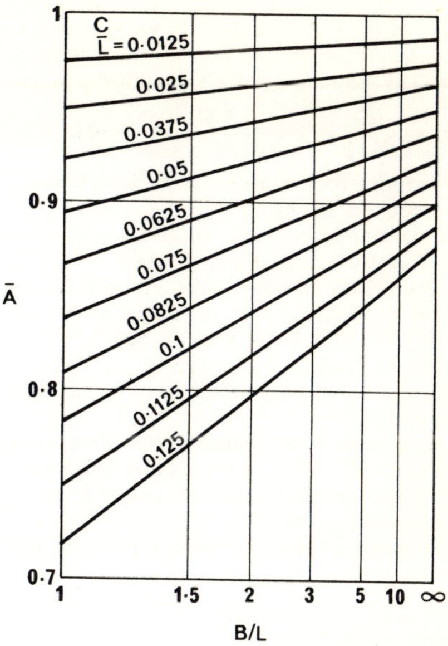

Fig 4

Flow parameter for rectangular multi-recess pads

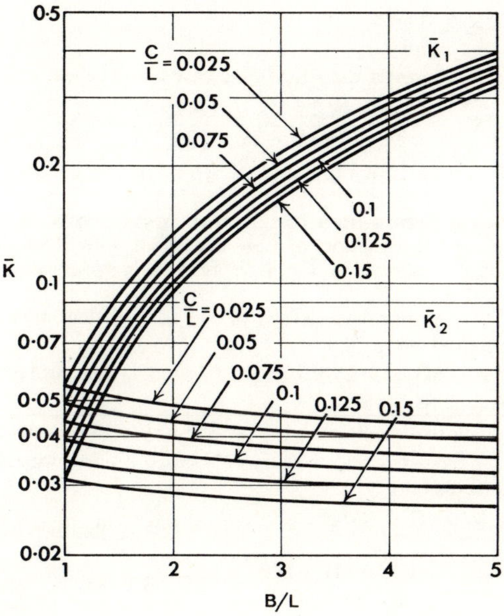

Fig 5

Tilting stiffness parameter for capillary controlled rectangular multi-recess pads

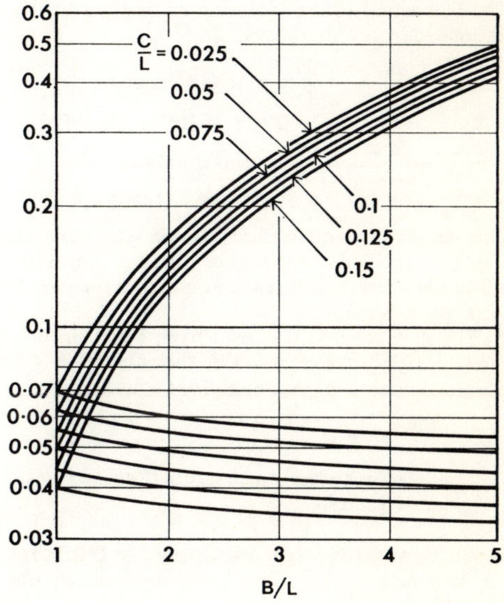

Fig 6

Tilting stiffness parameter for orifice controlled rectangular multi-recess pads

(2) The load coefficient is read from Fig 3, in this case with $B/L = 2{\cdot}67$, $C/L = 0{\cdot}125$, the value of $\bar{A}$ is 0·81.

(3) The minimum supply pressure is determined from equation (1).

$$P_s = \frac{W_o}{A\bar{A}\beta} = \frac{1{,}500}{6 \times 0{\cdot}81 \times 0{\cdot}5} = 610 \text{ lbf/in}^2$$

Note: The tilting parameter value $\bar{K}$ varies by only 10 per cent between the values of $\beta = 0{\cdot}35$ and $\beta = 0{\cdot}65$ for capillary compensation, and similarly only 10 per cent between the values of $\beta = 0{\cdot}45$ and 0·75 for orifice compensation. Beyond these limits, the stiffness and tilting stiffness both decrease. For example, in this case it is desired to use a supply pressure of 500 lbf/in² and since area is limited, then β is changed to 0·61. A reasonable estimation of the tilting stiffness could be made using Fig 5 or 6.

(4) The static stiffness parameter is determined from Fig 2.8 — 2.13. In this case we choose an orifice-controlled bearing with $\beta = 0{\cdot}61$, and read from Fig2 12 that $\bar{\lambda}_o = 1{\cdot}05$.

(5) The static stiffness is calculated using equation (3)

$$\lambda_o = \frac{P_s A}{h_o}\bar{A}\bar{\lambda}_o = \frac{500 \times 6}{0{\cdot}0015} \times 0{\cdot}81 \times 1{\cdot}05$$
$$= 1{\cdot}7 \times 10^6 \text{ lbf/in}$$

(6) The tilting parameter $\bar{K}$ for orifice control is read off for both axes from Fig 6, $\bar{K}_1 = 0{\cdot}18$, $\bar{K}_2 = 0{\cdot}036$ with $B/L = 2{\cdot}67$, and $C/L = 0{\cdot}125$.

(7a) The tilting stiffnesses for tilt about both lines of symmetry are calculated using equation 4

$$k_{o1} = \frac{P_s A L^2}{2h_o}\bar{K}_{o1} = \frac{500 \times 6 \times 1{\cdot}5^2}{0{\cdot}003} \times 0{\cdot}18$$
$$= 0{\cdot}405 \times 10^6 \text{ lbf.in/radian}$$
$$k_{o2} = 0{\cdot}81 \times 10^5 \text{ lbf.in/radian}$$

(7b) The actual value of maximum applied torque can be estimated from this stiffness

$$\theta_{max1} = \frac{2h_o}{B} = \frac{0{\cdot}003}{4} = 0{\cdot}00075 \text{ radians}$$
$$\therefore K_{1max} = 0{\cdot}405 \times 0{\cdot}00075 \times 10^6 = 304 \text{ lbf in}$$
$$\theta_{max2} = \frac{2h_o}{L} = \frac{0{\cdot}003}{1{\cdot}5} = 0{\cdot}002 \text{ radians}$$
$$\therefore K_{2max} = 0{\cdot}81 \times 10^5 \times 0{\cdot}002 = 162 \text{ lbf in}$$

This value represents a tilt condition where the edge of the pad is touching the runner surface. Since such a condition is extreme, safe maximum design torques should be *half* the above values.

(8) The flow coefficient is read from Fig 4, in this case for $B/L = 2{\cdot}67$ and $C/L = 0{\cdot}125$, the value of $\bar{B}$ is 4·5.

(9) The flow rate is calculated from equation (2)

$$q = \frac{P_s h_o^3}{\eta}\beta\bar{B} = \frac{500 \times (1{\cdot}5)^3 \times 10^{-9}}{25 \times 10^{-6}} \times 0{\cdot}61 \times 4{\cdot}5$$
$$= 0{\cdot}184 \text{ in/sec for one bearing (four recesses)}$$
$$= 0{\cdot}04 \text{ gal/min}$$

Design procedure for capillary restrictors

(1) Determine mass density of oil ρ in consistent units of lbf, in and sec

$$\rho = \text{S.G.} \times 9{\cdot}33 \times 10^{-5} \qquad (\text{lbf.in}^{-4}.\text{sec}^2)$$

For oil S.G. = 0·87

$$\rho = 0{\cdot}87 \times 9{\cdot}33 \times 10^{-5} = 8{\cdot}1 \times 10^{-5} \text{ lbf.in}^{-4}\text{ sec}^2$$

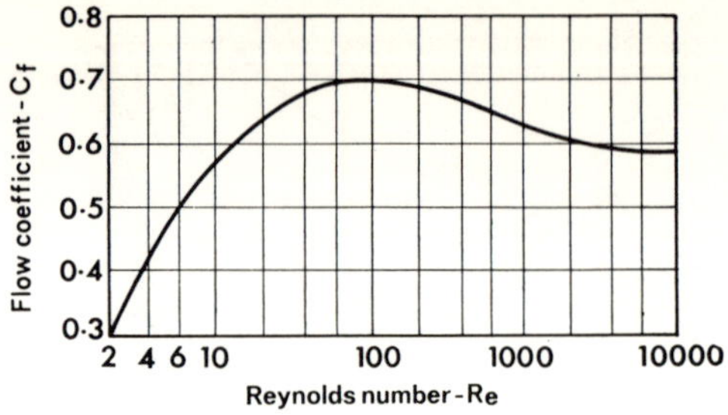

$\rho = 0{\cdot}000078$ to $0{\cdot}000083$ (lbf in sec units)

Fig 7

Curve relating Reynolds number to flow coefficient for a range of light-medium machine oils

(2) Calculate capillary constant K_c based on flow through one recess only

$$K_c = \frac{(1-\beta)P_s}{q\eta} \qquad (\text{in}^{-3})$$
$$= \frac{(1-0{\cdot}61) \times 500}{(0{\cdot}184 \div 4) \times 25 \times 10^{-6}} = 1{\cdot}7 \times 10^8 \quad (\text{in}^{-3})$$

(3) Calculate capillary bore d from $d^3 = \dfrac{128\,l}{\pi K_c d}$ for a suitable value of l/d preferably greater than 100 to eliminate effects of the entrance and exit conditions.

$$d^3 = \frac{128 \times 100}{\pi \times 1{\cdot}7 \times 10^8} \quad \text{for } l/d = 100$$
$$d = 0{\cdot}034 \text{ in} \qquad l = 3{\cdot}4 \text{ in}$$

(4) Calculate Reynolds number R_e

$$R_e = \frac{4\rho q}{\pi d \eta}$$

If R_e is greater than 1,000, increase d and repeat

$$R_e = \frac{4 \times 8{\cdot}1 \times 10^{-5} \times (0{\cdot}184 \div 4)}{\pi \times 0{\cdot}034 \times 25 \times 10^{-6}} = 5{\cdot}55$$

This value is very low, and there is no further concern.

Design procedure for orifice restrictors

A chart for variation of flow coefficient with Reynolds number is shown in Fig 7. It is initially assumed that $C_f = 0{\cdot}6$.

(1) Determine mass density of oil ρ in consistent units of lbf, in and sec

$$\rho = \text{SG} \times 9{\cdot}33 \times 10^{-5} \qquad (\text{lbf. in}^{-4}\text{ sec}^2)$$

For oil SG = 0·87

$$\rho = 0{\cdot}87 \times 9{\cdot}33 \times 10^{-5} = 8{\cdot}1 \times 10^{-5} (\text{lbf. in}^{-4}\text{ sec}^2)$$

(2) Calculate orifice constant K_{or} based on flow through one recess only

$$K_{or} = \frac{(1-\beta)P_s}{q_o^2} \qquad (\text{lbf. in}^{-8}\text{ sec}^2)$$
$$= \frac{(1-0{\cdot}61) \times 500}{(0{\cdot}184 \div 4)^2} = 9{\cdot}25 \times 10^4 (\text{lbf. in}^{-8}\text{ sec}^2)$$

(3) Calculate orifice area A

$$A = \frac{1}{C_f}\sqrt{\left(\frac{\rho}{K_{or}}\right)} \qquad (\text{in}^2)$$

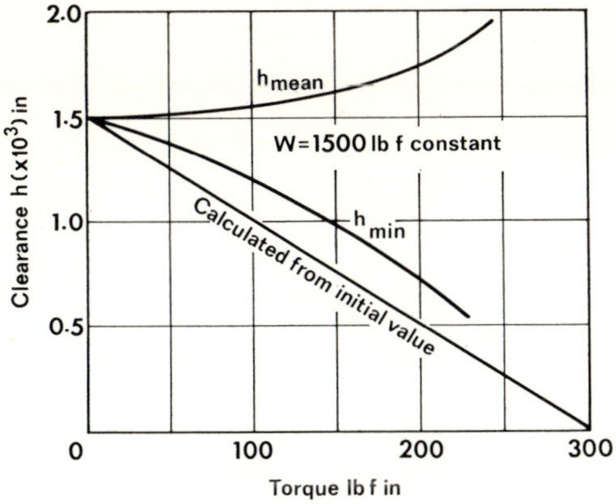

Fig 8

Effect of tilt on mean and minimum clearance for example given in appendix

Assume $C_f = 0{\cdot}6$ unless more accurately known

$$A = \frac{1}{0{\cdot}6}\sqrt{\frac{8{\cdot}1 \times 10^{-5}}{9{\cdot}25 \times 10^{4}}} = 1{\cdot}53 \times 10^{-4} \qquad (\text{in}^2)$$

(4) Calculate orifice diameter d

$$d = \sqrt{\frac{4A}{\pi}} \qquad (\text{in})$$

$$d = \sqrt{\frac{4 \times 1{\cdot}53 \times 10^{-4}}{\pi}} = 0{\cdot}014 \qquad (\text{in})$$

(5) Calculate Reynolds number $R_e = \dfrac{4\rho q}{\pi d \eta}$

$$R_e = \frac{4 \times 8{\cdot}1 \times 10^{-5} \times (0{\cdot}184 \div 4)}{\pi \times 0{\cdot}014 \times 25 \times 10^{-6}} = 13{\cdot}5$$

Checking with Fig 7, this gives $C_f = 0{\cdot}6$ and there is no need to repeat the procedure.

High-speed applications

Normally these bearings are not used for applications where there is high relative speed, and if there is a high relative speed, the tilting torque stiffness will be seriously influenced by hydrodynamic effects and entrained velocity of fluid into or out of the pockets. If there is a tilt such that the clearance of the leading edge of the bearing is greater than the mean clearance, the tilting stiffness will be increased and the bearing will tend to lift, whereas if the clearance of the leading edge is lower than the mean clearance, the tilting stiffness will be reduced and the bearing may collapse. This problem makes essential the use of viscosities and clearances such that the optimization limit is not exceeded.

The optimization limit used by the authors is that the friction loss should not exceed the pumping power loss. Using the above example the limiting speed can be calculated thus:

(9) Pumping power loss $H_p = P_s \,.\, q = 500 \times 0{\cdot}1864 = 92{\cdot}5$ lbf. in/s

(10) Recess area $= 0{\cdot}375 \times 1{\cdot}625 \times 4 = 2{\cdot}44$ in^2

(11) Effective friction area = land area + $\frac{1}{4}$ recess area

$$A_f = \text{total area} - \tfrac{3}{4}\text{ recess area}$$
$$= 6 - \tfrac{3}{4} \times 2{\cdot}44$$
$$= 6 - 1{\cdot}83 = 4{\cdot}17 \text{ in}^2$$

(12) Friction power loss $H_f = \eta A_f \dfrac{V^2}{h_o}$

$$= 25 \times 10^{-6} \times 4{\cdot}17 \times \frac{V^2}{0{\cdot}0015}$$
$$= 69{\cdot}4V^2 \ \frac{\text{lbf in}}{s}$$

(13) The critical velocity therefore is given by

$$69{\cdot}4\ V_c^2 = 92{\cdot}5$$
$$V_c = 1{\cdot}15 \text{ in/s}$$

If the maximum speed is known to be, say, 10 in/s then the viscosity should be reduced thus

$$\eta_{new} = \eta \,.\, \frac{V_c}{V} = 25 \times 10^{-6}\,\frac{1{\cdot}15}{10}$$
$$= 2{\cdot}88 \times 10^{-6} \text{ reyn.}$$

Appendix

Behaviour of a tilted multi-recess bearing under varying load conditions

The above design procedure gives the stiffness and tilting resistance of a multi-recess bearing at the design condition. The behaviour of this type of bearing at conditions other than the design condition is complex. As the load is increased, the effective value of the pocket pressure ratio increases and the value of $\bar{K}$ will be altered. Similarly the effect of large tilting moments is to alter the values of the pocket pressures unevenly, and the mean film thickness for constant load increases. To illustrate this behaviour, the computed performance of the bearing designed in the above case study is given in Fig 8 which shows the variation of film thickness with applied tilting moment for a constant value of load. The mean film thickness actually increases with applied torque, but the edge film thickness lessens. The flow rate is reduced by approximately 6 per cent at a tilting torque of 200 lbf.in.

The computed results indicate three important points for the designer.

(1) The tilting stiffness at the design condition is conservative due to the increase of load coefficient with tilt which causes an increase in mean film thickness.

(2) The flow at the design condition is also a good design figure since tilt tends to reduce the flow.

(3) The tilting stiffness decreases with tilt and although conservative at the design condition it will be unreliable for a tilt such that the minimum clearance is less than 25 per cent of the mean.

REFERENCES

1. J P O'Donoghue and W. B. Rowe. "Hydrostatic Bearing Design." *Tribology*, Vol 2, No 1, February 1969, pp. 25-71.

Chapter Seven

Spherical bearings

J P O'DONOGHUE

Spherical thrust bearings such as those shown in Fig 1 are used to carry loads in situations where it is important to have very low resistance to tilting moments, as required for pivots and for self-aligning bearings, for example. They can be simple single pads with a central recess, Fig 1a, or with an annular recess, Fig 1b, or they may be multi-recess bearings used either singly or as opposed pairs, as in Fig 1c and 1d. In this chapter, as in previous chapters, the aim has been to present a straightforward design procedure which will permit rapid calculation of load, stiffness, and flow for the type of bearing under discussion. The method is a step-by-step procedure using design charts which provide the data to enable bearing performance to be calculated from simple basic relationships.

Basic relationships

The basic load equation is

$$W_o = P_s A \bar{A}_o \beta \quad \text{............} \quad (1)$$

The basic flow equation is

$$q_o = \frac{P_s C^3}{\eta} \beta \bar{B}_o \quad \text{............} \quad (2)$$

The basic stiffness equation is

$$\lambda_o = \frac{P_s A \bar{\lambda} \bar{A}_o}{C} \quad \text{................} \quad (3)$$

The stiffness values throughout are for capillary-controlled bearings, but a useful estimate for orifice-controlled bearings is to consider the concentric stiffness as being 33 per cent higher than that for capillary-controlled bearings. The collapse load for the simple bearing is slightly reduced when it is of the orifice-controlled type.

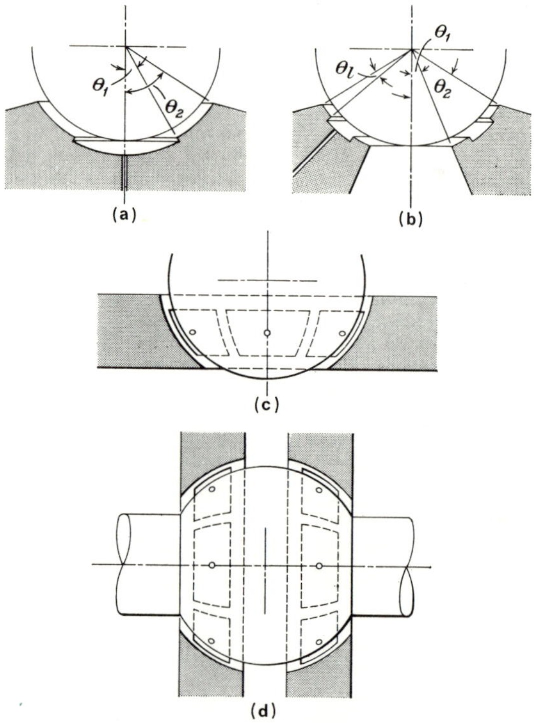

Fig 1

Various types of spherical hydrostatic bearings are here shown, namely (a) single central recess; (b) single annular recess; (c) multi-recess; and (d) opposed multi-recess

Symbols

Suffix $_o$ Denotes concentric condition
- P_s Supply pressure
- P Recess pressure
- W Bearing load (radial load in multi-recess bearing)
- T Thrust load in multi-recess bearing
- A Area
- $\bar{A}$ Area factor
- $\varepsilon = 1 - h_{min}/C$
- A_a Axial projected area
- A_r Radial projected area
- β P_o/P_s Design pressure ratio
- C Radial clearance
- $\bar{B}$ Flow factor
- η Viscosity (absolute)
- q Flow rate
- λ Viscosity
- γ Ratio θ_1/θ_2 (see Fig 1)
- D Diameter of sphere

Simple single-recess bearing

The simple single-recess bearing has a disadvantage in that it will collapse at high load and exhibit negative stiffness.[1] However, if the load is constant, and never exceeds a critical value, this simple geometry can be used. Load, flow and stiffness values are given in Fig 2 to 4. The recommended value of the pressure ratio β is 0·5, and the land-width angular ratio, γ, 0·5. Use of the charts is illustrated in the following example.

A thrust load of 240 lbf is to be carried by a single-recess spherical bearing, and the design steps are as follows.

(1) The design load is calculated; $W_o = 240$ lbf.
(2) The supply pressure is chosen; $P_s = 300$ lbf/in^2
(3) The angular land-width ratio is initially taken as $\gamma = 0{\cdot}5$
(4) The value of $\bar{A}_o$ is assumed; $\bar{A}_o = 0{\cdot}6$
(5) The diameter of the sphere is chosen; $D = 2$ in
(6) The bearing area is calculated from

$$A = \frac{W_o}{P_s \bar{A}_o \beta} = \frac{240}{300 \times 0{\cdot}6 \times 0{\cdot}5}$$
$$= 2{\cdot}67 \text{ in}^2$$

(7) The bearing angle is calculated from

$$\sin^2 \theta_2 = \frac{4A}{\pi D^2} = \frac{4 \times 2{\cdot}67}{\pi \times 4} = 0{\cdot}85$$
$$\sin \theta_2 = 0{\cdot}92$$
$$\therefore \theta_2 = 67 \text{ deg}$$

(8) The land angle $\theta_1 = \theta_2 \times \gamma = 67 \times 0{\cdot}5 = 33{\cdot}5$

NOTE: Steps 5, 6, 7 and 8 can be rearranged if it is desired to use specific values of θ_2 and θ_1 and to calculate the necessary diameter, thus

(5_1) θ_2 is taken as 60 deg
(6_1) The area is calculated as before $A = 2{\cdot}67$ in^2
(7_1) The sphere diameter is calculated from

$$D = \frac{2}{\sin \theta_2}\sqrt{\frac{A}{\pi}} = \frac{2}{0{\cdot}866}\sqrt{\frac{2{\cdot}67}{\pi}}$$
$$= 2{\cdot}13 \text{ in}$$

(8_1) The land angle is given by $\theta_1 = \theta_2 \times \gamma$
$= 60 \times 0{\cdot}5 = 30$ deg

(9) Concentric load capacity is calculated, after reading $\bar{A}_o$ from Fig. 2.

For $\theta_2 = 67$ deg and $\gamma = 0{\cdot}5$, $\bar{A}_o$ is read as 0·67.

$\therefore W = P_s \beta A \bar{A}_o = 300 \times 0{\cdot}5 \times 2{\cdot}67 \times 0{\cdot}67$
$= 267$ lbf.

This small discrepancy is due to the assumed value of $\bar{A}_o = 0{\cdot}6$ in step 4, before the angle θ_2 was known. At this point, the designer may choose to recalculate from step 4, or he can alter β to a smaller value, or simply be content to operate at a slightly positive value of eccentricity (in this case it has been computed as 0·15).

(10) The stiffness parameter is read from Fig 4. For convenience, this value is combined with the load parameter.

For $\theta_2 = 67$ deg and $\gamma = 0{\cdot}5$
$\bar{\lambda} \bar{A} = 0{\cdot}28$

(11) The clearance is chosen; $C = 0{\cdot}001$ in
(12) The stiffness is calculated from

$$\lambda_o = \frac{P_s A}{C} \bar{\lambda} \bar{A}$$

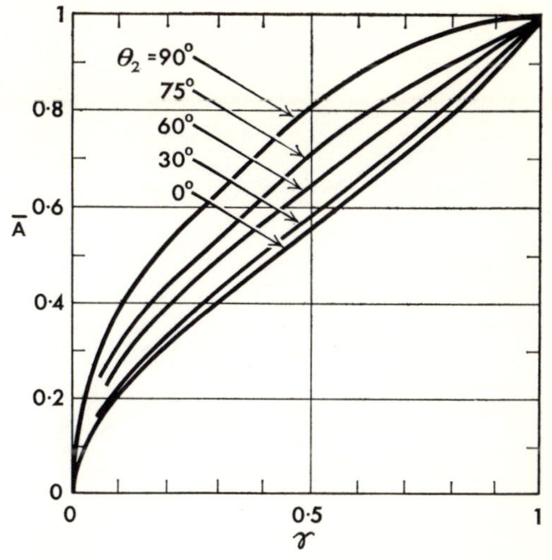

Fig 2

Curves for determining the area factor $\bar{A}$ for a single central recess bearing

Fig 3

Curves for determining the flow factor $\bar{B}_o$ for a single central recess bearing

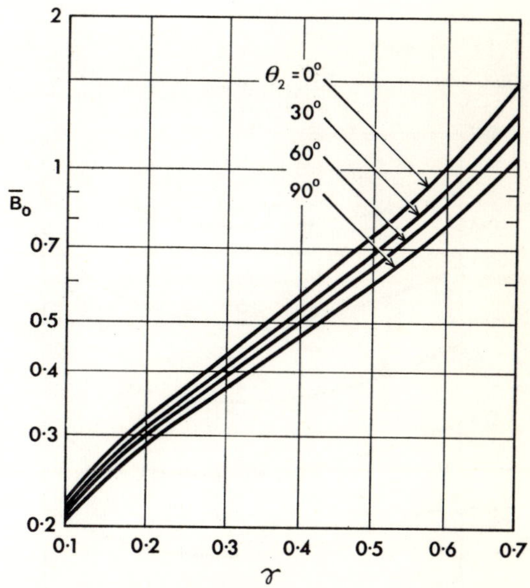

$$= \frac{300 \times 2{\cdot}67}{10^{-3}} \times 0{\cdot}28$$
$$= 2{\cdot}24 \times 10^5 \text{ lbf/in}$$

(13) Viscosity (absolute) is chosen; $\eta = 30$cp

$$= 30 \times 1{\cdot}45 \times 10^{-7} = 4{\cdot}35 \times 10^{-6} \frac{\text{lbf . s.}}{\text{in}^2}$$

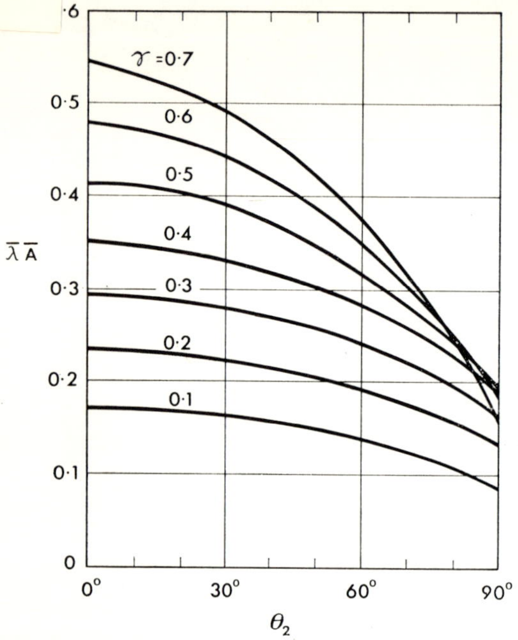

Fig 4

Curves for determining the stiffness factor $\bar{\lambda}\ \bar{A}$ for a single central recess bearing

Fig 5

Maximum load curves for a single central recess bearing

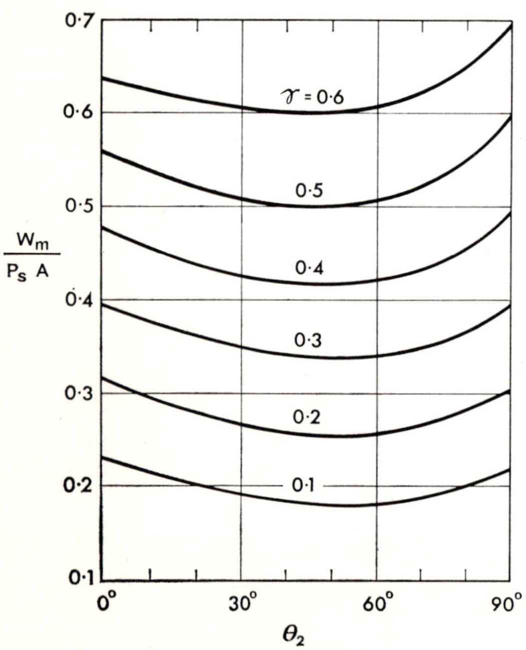

(14) Flow factor is read from Fig 3;

For $\theta_2 = 67$ deg, $\gamma = 0{\cdot}5$

$\bar{B}_o = 0{\cdot}665$

NOTE: An exact expression is

$$\bar{B}_o = \frac{\pi}{6 \log_e \left\{\frac{\tan(\theta_2/_2)}{\tan(\theta_1/_2)}\right\}}$$

(15) The flow is calculated;

$$q_o = \frac{P_s\, C^3}{\eta}\ \beta\ \bar{B}$$

$$= \frac{300 \times 10^{-9}}{4{\cdot}35 \times 10^{-6}} \times 0{\cdot}5 \times 0{\cdot}665$$

$$= 2{\cdot}29 \times 10^{-2}\ \text{in}^3/\text{s}$$

(16) Conversion to convenient flow units;

$$q_o = 2{\cdot}29 \times 10^{-2} \times \frac{60}{277}$$

$$= 4{\cdot}96 \times 10^{-3}\ \text{gal/min}$$

(17) Check on maximum load.
Fig. 5 shows the maximum value of load. For this case

$$\frac{W_M}{P_s\ A} = 0{\cdot}52$$

$$W_M = 0{\cdot}52 \times 300 \times 2{\cdot}67 = 410\ \text{lbf.}$$

Annular thrust bearing

The annular thrust bearing has a number of advantages:

(1) The pad load coefficient is sensibly constant for varying eccentricity and therefore there is no problem with collapse.

(2) The shaft may be a through shaft arrangement with a reduced diameter, or an opposed pair of bearings can be used.

(3) By dividing the annular recess into several pockets, a side thrust can be accommodated.

The load, flow and stiffness coefficients are given in Fig 6 to 8.

Example: An annular spherical thrust bearing is to carry a load of 1,000 lbf. The through shaft diameter is 3 in. Determine a suitable geometry if the supply pressure is 300 lbf/in², the oil viscosity is 30 cp, and the clearance is 0·002 in.

The recommended values of $\theta_l = 0{\cdot}25\ (\theta_2 - \theta_1)$, and $\beta = 0{\cdot}5$ are chosen. For convenience, define the land

ratio, $\bar{c}, = \frac{\theta_l}{\theta_2 - \theta_1}$

(1) The design load is calculated; $W_o = 1{,}000$ lbf

(2) The supply pressure is chosen; $P_s = 300$ lbf/in²

(3) The load parameter is estimated as $1 - \bar{c}$; thus, $\bar{A}_o = 0{\cdot}75$

(4) The annular area is calculated;

$$A = \frac{W}{P_s\ \beta\ \bar{A}_o} = \frac{1{,}000}{300 \times 0{\cdot}5 \times 0{\cdot}75}$$

$$= 8{\cdot}9\ \text{in}^2$$

(5) The outer diameter is calculated;

$$D_o = \sqrt{\frac{4A}{\pi} + D_i^2}$$

The inner diameter, $D_i = 3{\cdot}125$ in, ie 3 in + 0·125 in clearance, so that;

$$D_o = \sqrt{\frac{4 \times 8{\cdot}9}{\pi} + 3{\cdot}125^2}$$
$$= 4{\cdot}6 \text{ in}$$

(6) The designer can now decide on the angles. The reason for using a spherical bearing as opposed to a plane annular bearing is the advantage of alignment and the shaft will have virtually zero resistance to tilt about a point at the centre of the spherical surface. The location of this centre should be carefully considered by the designer. In this example, it is decided to locate the centre of the sphere at a point 3 in along the shaft from the thrust bearing surface, this point being the location of a cylindrical journal bearing. The angles θ_2 and θ_1 are thus calculated as follows:

$$\theta_1 = \sin^{-1}\frac{3{\cdot}125}{6} = 31{\cdot}5 \text{ deg}$$

$$\theta_2 = \sin^{-1}\frac{4{\cdot}60}{6} = 50 \text{ deg}$$

$$\gamma = \theta_1/\theta_2 = \frac{31{\cdot}5}{50} = 0{\cdot}63$$

(7) The flow factor is read from Fig 7;

For $\theta_2 = 50$ deg and $\gamma = 0{\cdot}63$
$\bar{B}_o = 7{\cdot}7$

(8) The flow may be calculated from

$$q = \frac{P_s C^3}{\eta}\, \beta\, \bar{B}_o$$
$$= \frac{300 \times (0{\cdot}002)^3}{30 \times 1{\cdot}45 \times 10^{-7}}\, 0{\cdot}5 \times 7{\cdot}7$$
$$= 2{\cdot}13 \text{ in}^3/\text{s}$$
$$= 0{\cdot}46 \text{ gal/min}$$

(9) The stiffness parameter is read from Fig 8. Again, for convenience, $\bar{\lambda}\bar{A}$ has been given. In this case

$\theta_2 = 50$ deg, $\gamma = 0{\cdot}63$
$\bar{\lambda}\bar{A} = 0{\cdot}42$

(10) The stiffness is therefore

$$\lambda = \frac{P_s A\, \bar{\lambda}\bar{A}}{C} = \frac{300 \times 8{\cdot}9}{0{\cdot}002} \times 0{\cdot}42 = 5{\cdot}6 \times 10^5 \text{ lbf/in}$$

(11) The concentric load is re-calculated using the correct value for $\bar{A}_o$ from Fig 6. In this case,

$\theta_2 = 50$ deg, $\gamma = 0{\cdot}63$, $\bar{A}_o = 0{\cdot}76$

so that, $W_o = 300 \times 8{\cdot}9 \times 0{\cdot}76 \times 0{\cdot}5 = 1{,}010$ lbf

Approximate method for spherical multi-recess bearings

Although the author has analysed the spherical multi-recess bearing and produced a general analysis computer program, the number of variables is too great to permit a full presentation of the computed results. The author is prepared to assist industry by running the program as a design service without charge.

The results obtained from this program have been used to check the following approximate design procedure. The author has based the procedure on a maximum design load which will give an eccentricity of 0·5. The line of action of the load may be assumed to pass through the

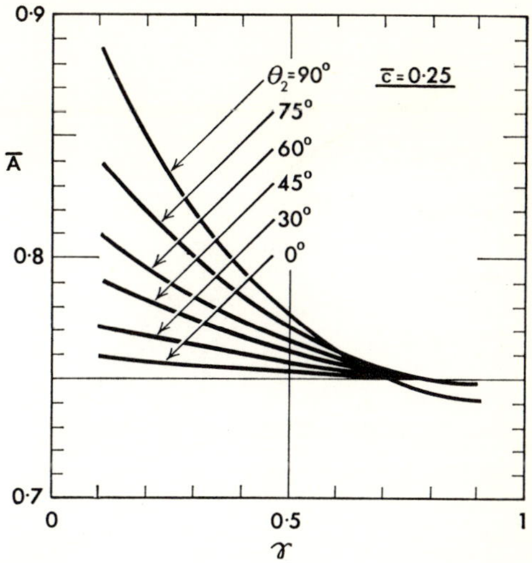

Fig 6

Curves for determining the area factor $\bar{A}$ for an annular recess bearing

Fig 7

Curves for determining the flow factor $\bar{B}_o$ for an annular recess bearing

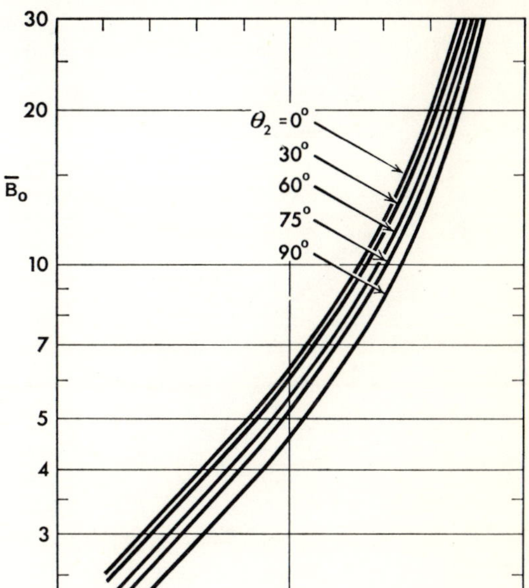

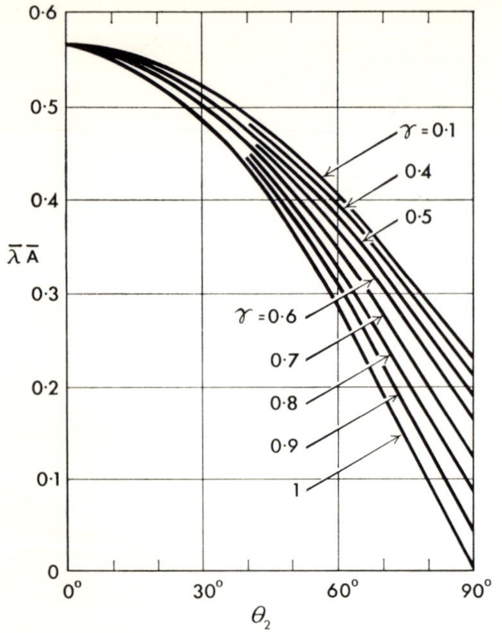

Fig 8

Curves for determining the stiffness factor $\bar{\lambda}$ $\bar{A}$ for an annular recess bearing

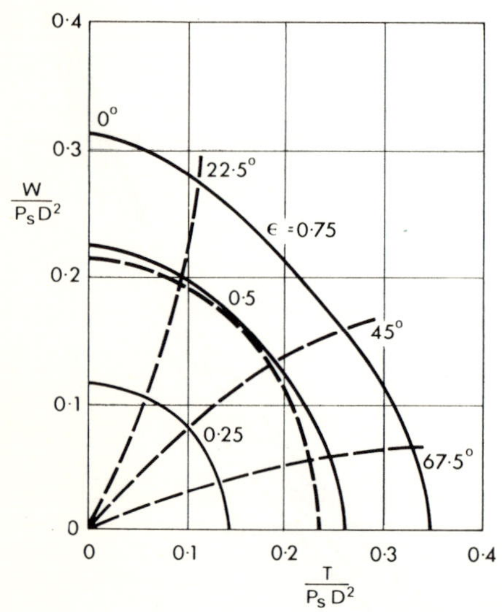

Fig 9

Radial load W and direct thrust T for a multi-recess bearing

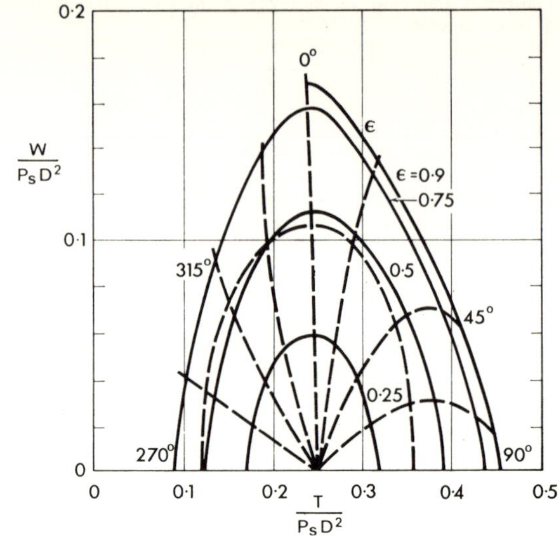

Fig 10

Radial load W and direct thrust T for an opposed pair of multi-recess bearings

centre of the sphere. As with a conical bearing, chapter three, the single spherical recess bearing will have a radial load capacity dependent on the axial load.

Unlike previous design calculations this method involves determination of the performance of a given bearing geometry rather than the converse approach of determining an optimum geometry for a given performance. However the basic optimization assumptions of pressure ratio $\beta = 0{\cdot}5$ and land-width ratio $c = 0{\cdot}25$ remain the same. It is recommended that a 6-pocket geometry is used. By the nature of a spherical bearing, the length to diameter ratio must be less than 0·5, and a 4-pocket arrangement consequently gives a poor performance. The angle of the interpocket land is made equal to the axial land angle.

Example: A 6-pocket bearing is designed with $\theta_2 = 88$ deg, $\theta_1 = 30$ deg, $D = 3{\cdot}5$ in, $P_s = 300$ lbf/in², $c = 0{\cdot}25$. Determine the characteristics of the bearing used singly and in an opposed-pair arrangement.

(1) Major angle; $\theta_2 = 88$ deg.

(2) Minor angle; $\theta_1 = 30$ deg.

(3) Land-width ratio; $\bar{c} = 0{\cdot}25$

(4) Land-width angle, $\theta_l = (88 - 30)0{\cdot}25 = 7$ deg

(5) The axial and radial projected area ratios are now calculated;

$$\frac{A_a}{D^2} = \frac{\pi}{4}\left(\sin^2\theta_2 - \sin^2\theta_1\right)$$

$$= \frac{\pi}{4}\left(F_2(\theta_2) - F_2(\theta_1)\right)$$

$$= \frac{\pi}{4}\left(0{\cdot}999 - 0{\cdot}25\right) = 0{\cdot}59$$

$$\frac{A_r}{D^2} = \frac{1}{4}\left(\theta_2 - \frac{\sin 2\theta_2}{2} - \left(\theta_1 - \frac{\sin 2\,\theta_1}{2}\right)\right)$$

$$= \frac{1}{4}\left(F_1(\theta_2) - F_1(\theta_1)\right)$$

$$= \frac{1}{4}\left(1{\cdot}501 - 0{\cdot}0906\right)$$

$$= 0{\cdot}353$$

For convenience, values of F_2, $(\sin^2 \theta)$, and F_1, $\left(\theta - \frac{\sin 2\theta}{2}\right)$ for various values of the angles concerned, are given in the accompanying tables.

Values for the factors F_1 and F_2 for different angles θ_1 and θ_2, to facilitate calculation of axial and radial projected area ratios

θ	$F_1(\theta)$	$F_2(\theta)$	θ	$F_1(\theta)$	$F_2(\theta)$
1	0·00000	0·00030	46	0·30316	0·51745
2	0·00003	0·00122	47	0·32152	0·53488
3	0·00010	0·00274	48	0·34050	0·55226
4	0·00023	0·00487	49	0·36008	0·56959
5	0·00044	0·00760	50	0·38026	0·58682
6	0·00076	0·01093	51	0·40104	0·60396
7	0·00121	0·01485	52	0·42242	0·62096
8	0·00181	0·01937	53	0·44439	0·63782
9	0·00257	0·02447	54	0·46695	0·65451
10	0·00352	0·03015	55	0·49008	0·67101
11	0·00468	0·03641	56	0·51379	0·68730
12	0·00607	0·04323	57	0·53806	0·70337
13	0·00771	0·05060	58	0·56289	0·71919
14	0·00961	0·05853	59	0·58827	0·73474
15	0·01180	0·06699	60	0·61418	0·75000
16	0·01429	0·07598	61	0·64063	0·76496
17	0·01711	0·08548	62	0·66759	0·77960
18	0·02027	0·09549	63	0·69505	0·79389
19	0·02378	0·10599	64	0·72301	0·80783
20	0·02767	0·11698	65	0·75144	0·82139
21	0·03195	0·12843	66	0·78034	0·83457
22	0·03664	0·14033	67	0·80970	0·84733
23	0·04176	0·15267	68	0·83949	0·85967
24	0·04731	0·16543	69	0·86971	0·87157
25	0·05331	0·17861	70	0·90034	0·88302
26	0·05978	0·19217	71	0·93135	0·89401
27	0·06673	0·20611	72	0·96274	0·90451
28	0·07417	0·22040	73	0·99449	0·91452
29	0·08212	0·23504	74	1·02658	0·92402
30	0·09059	0·25000	75	1·05900	0·93301
31	0·09958	0·26526	76	1·09171	0·94147
32	0·10911	0·28081	77	1·12472	0·94940
33	0·11919	0·29663	78	1·15799	0·95677
34	0·12982	0·31270	79	1·19151	0·96359
35	0·14102	0·32899	80	1·22525	0·96985
36	0·15279	0·34549	81	1·25921	0·97553
37	0·16514	0·36218	82	1·29335	0·98063
38	0·17808	0·37904	83	1·32766	0·98515
39	0·19160	0·39604	84	1·36212	0·98907
40	0·20573	0·41318	85	1·39671	0·99240
41	0·22045	0·43041	86	1·43140	0·99513
42	0·23578	0·44774	87	1·46617	0·99726
43	0·25171	0·46512	88	1·50101	0·99878
44	0·26825	0·48255	89	1·53589	0·99970
45	0·28540	0·50000	90	1·57080	1·00000

(6) The radial load at the designed thrust is calculated;

$$W_o = 0{\cdot}4\,(1 - \bar{c})\,P_s\,A_r$$
$$= 0{\cdot}4\,(1 - 0{\cdot}25) \times 300 \times 0{\cdot}353 \times 3{\cdot}5^2$$
$$= 391 \text{ lbf}$$

(7) The axial value of $\bar{A}$ is read from Fig 6.
For this case $\bar{A} = 0{\cdot}81$ ($\theta_2 = 88$ deg and $\gamma = 0{\cdot}342$).

(8) The concentric axial thrust load is calculated.

$$T_o = 0{\cdot}5\,P_s\,A_a\,\bar{A}$$
$$= 0{\cdot}5 \times 300 \times 0{\cdot}59 \times 3{\cdot}5^2 \times 0{\cdot}81$$
$$= 878 \text{ lbf}$$

The approximate single bearing characteristic is given by an ellipse with a centre at $W = 0$, $T_o = 878$ lbf, and the axes have co-ordinates $W = 0$, $T = \frac{3T_o}{2}$ and $\frac{T_o}{2}$, and $T = T_o$, $W = \pm W_o$. This characteristic is shown, together with the computed characteristic, in Fig. 9. As a conservative estimate this procedure has proved a good prediction for this example.

The approximate opposed bearing characteristic is given by an ellipse with a centre at $W = 0$, $T = 0$, and the axes are $W = 0$, $T = \pm T_o$ and $T = 0$, $W = 2W$. This characteristic is shown in Fig 10, together with the computed performance. Again, the agreement is good. The procedure as regards factors related to flow is the same as for a single annular recess case, but the total flow is divided by the number of recesses before the capillary or orifice control dimensions are calculated.

REFERENCES

1. J P O'Donoghue and G K Lewis. 'Spherical hydrostatic bearings', *Tribology*, Vol. 4, No. 3, November 1970.

Chapter Eight

Annular multi-recess thrust bearings

J P O'DONOGHUE

Annular thrust bearings of the type shown in Fig 1 are used for carrying an axial load on a shaft collar, or as support bearings for turntables. They may be used for uni-directional loads or for reversing loads, the latter requiring an opposed-pad arrangement.

In this chapter as in earlier chapters, the aim is to present a straightforward design procedure which will permit the rapid calculation of load, stiffness, flow and tilt resistance for the type of bearing under discussion. The method is a step-by-step procedure, using design charts which provide the data necessary to enable bearing performance to be calculated using simple basic relationships.

Annular bearing arrangements for shaft collars

There are several arrangements for collar bearings, the two principal types being shown in Fig 1 at (*a*) and (*b*). The multi-recess bearing is in general preferable to the multi-pad arrangement since it has two advantages:

(1) the thrust load capacity is higher for a given supply pressure, and

(2) the flow is reduced.

These factors are considered sufficient to offset the marginally lower tilt resistance of a multi-recess bearing compared with a multi-pad type.

The number of recesses is taken to be six, which is a compromise between high tilt capacity and low directional variations of tilt resistance on the one hand and the cost of machining and the complexity of supply systems on the other. Each recess has its own feed and control device, which may be either capillary or orifice, and as in the previous procedures a value of 0·5 is recommended for the ratio of design recess pressure to supply pressure, β, to provide high load and stiffness to minimise the effects of bearing manufacturing tolerances and to give a reasonably-sized bearing.

The land-width between the recesses should be twice the width of the radial flow lands, and the ratio of the latter to the annular bearing width should be taken as 0·25.

The load relationship is given by

$$W_o = P_s A \bar{A} \beta \quad \ldots\ldots\ldots\ldots\ldots \quad (1)$$

and in this instance

$$A = \frac{\pi}{4}(D_o^2 - D_i^2)$$

The flow rate is expressed

$$q_o = \frac{P_s h_o^3}{\eta} \beta \bar{B}_o \quad \ldots\ldots\ldots\ldots\ldots \quad (2)$$

The stiffness is expressed

$$\lambda = \frac{P_s A \bar{A}}{h_o} \bar{\lambda} \quad \ldots\ldots\ldots\ldots\ldots \quad (3)$$

and the tilting stiffness is expressed

$$k_o = \frac{P_s A D_o^2}{2h_o} \bar{K} \quad \ldots\ldots\ldots\ldots \quad (4)$$

The values of $\bar{A}$, $\bar{B}$, $\bar{\lambda}$ and $\bar{K}$ are given in Fig 2 to 5, the values for $\bar{K}$ being for capillary control only. The values of $\bar{K}$ for orifice control will be approximately

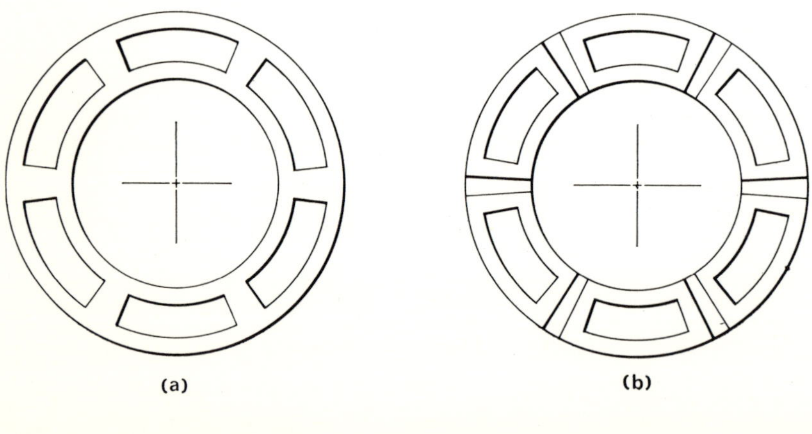

(a) (b)

Fig 1

Two types of annular hydrostatic thrust bearings, the multi-recess shown at (a) and the multi-pad shown at (b)

25 per cent higher, but although the parallel tilting stiffness value is increased, the actual limit to the applied tilt will be the same as for capillary control, so that even for orifice control it is suggested that the capillary value is used for determining this value.

Design procedure

The design procedure is best illustrated by a. numerical example. As in previous chapters, the slow-speed case is dealt with first and then the effect of speed is considered. In this example, a single annular multi-recess bearing for a 2·5-in diameter shaft is to carry a load of 1,000 lbf. The oil viscosity is taken as 25 × 10^{-6} reyn, and the supply pressure, as 300 lbf/in². A capillary control system is used.

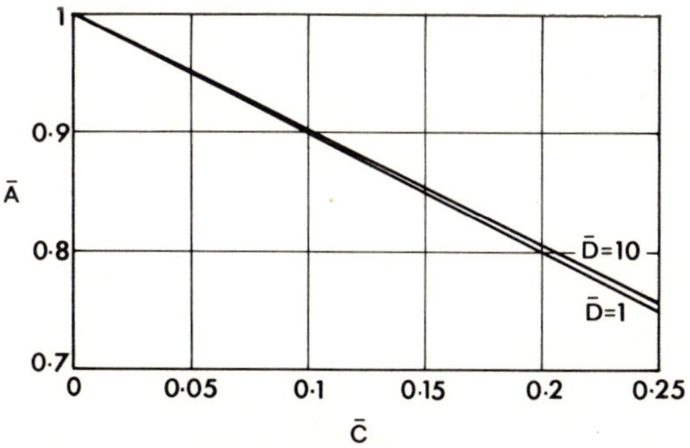

Note: Approximate formula $\bar{A} = 1 - \bar{C}$

Fig 2

Values for $\bar{A}$, the load factor, can be taken from this graph

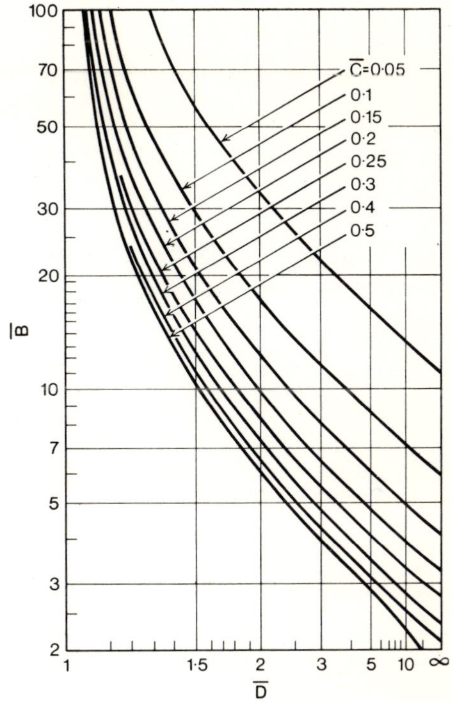

Fig 3

Curves for determining the flow shape factor $\bar{B}$

Symbols

- P_s — Supply pressure
- $\bar{P} = \dfrac{P_r}{P_s}$ — Ratio of recess pressure to supply pressure
- W — Supported load on bearing thrust
- A — Total projected bearing area
- A_e — Effective area over which recess pressure may be assumed to act in calculating W
- $\bar{A} = \dfrac{A_e}{A}$ — Area shape factor (or load factor) for bearing pad
- D_o — Outside bearing diameter
- D_i — Inside bearing diameter
- c — Radial land-width from recess to outside
- $\bar{c}$ — Land-width ratio $= \dfrac{2c}{D_o - D_i}$
- $\bar{D} = \dfrac{D_o}{D_i}$ — Diameter ratio of bearing
- $\bar{B}$ — Flow shape factor for bearing pad $= \dfrac{q\eta}{P_r h_o^3}$
- q — Flow rate
- q' — Flow for each pocket $= \dfrac{q_o}{n}$
- η — Dynamic viscosity
- h — Film thickness
- $_o$ — Suffix denoting designed operating condition
- λ — Stiffness
- $\bar{\lambda}$ — Dimensionless stiffness $= \dfrac{\lambda h_o}{P_s A \bar{A}}$
- k — Tilting stiffness
- $\bar{K}$ — Dimensionless tilting stiffness $= \dfrac{2k\,h}{P_s A D_o^2}$
- β — Design pressure ratio $= \dfrac{P_{ro}}{P_s}$
- θ — Angle of bearing tilt
- M — Tilting load on bearing
- $\bar{X} = \dfrac{h}{h_o}$ — Dimensionless film thickness
- H_p, H_f — Pumping power and friction power
- S_H — Optimization parameter $= \dfrac{\eta N}{P_s}\left(\dfrac{D_o}{h_o}\right)^2$
- ΔT — Temperature rise from oil inlet to oil outlet
- N — Rotational speed
- V — Sliding velocity
- A_f — Effective friction area

(1) The design load is calculated = 1,000 lbf
(2) The supply pressure is chosen = 300 lbf/in^2
(3) The land-width ratio is chosen initially as $\bar{c} = 0{\cdot}25$
(4) The value of $\bar{A}$ is read from Fig 2 for a value of $\bar{c} = 0{\cdot}25$. $\bar{A} = 0{\cdot}75$
(5) The pressure ratio is chosen initially as $\beta = 0{\cdot}5$
(6) The required area is calculated from equation 1

$$\bar{A} = \frac{W}{P_s \beta \bar{A}} = \frac{1000}{300 \times 0{\cdot}5 \times 0{\cdot}75} = 9{\cdot}9 \text{ in}^2$$

(7) The internal diameter is chosen, leaving a clearance between the bearing and the shaft. $D_i = 2{\cdot}75$ in
(8) The external diameter is calculated using the formula

Fig 4

The stiffness factor $\bar{\lambda}$ for a capillary-controlled bearing is taken from the curves (a) and for an orifice-controlled bearing from the curves (b)

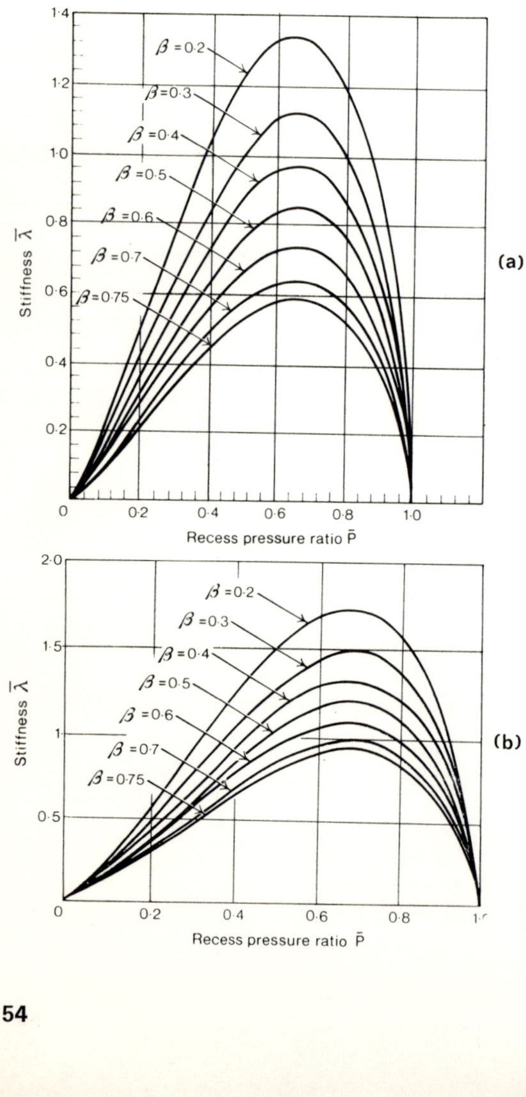

$$D_o = \sqrt{\frac{4A}{\pi} + D_i^2}$$
$$= \sqrt{11{\cdot}32 + 7{\cdot}56} = 4{\cdot}34 \text{ in}$$

(9) The diametral ratio is calculated. $\bar{D} = D_o/D_i = 1{\cdot}58$
(10) The value of $\bar{B}$ when $\bar{D} = 1{\cdot}58$ and $\bar{c} = 0{\cdot}25$ is read from Fig 3. $\bar{B} = 12{\cdot}5$
(11) The design clearance is chosen and is based on the size, surface finish and flatness that can be achieved. In this case, h_o is taken as 0·002 in.
(12) The flow is calculated using equation 2

$$q_o = \frac{P_s h_o^3 \beta \bar{B}}{\eta} = \frac{300 \times 8 \times 10^{-9}}{25 \times 10^{-6}} \times 0{\cdot}5 \times 12{\cdot}5 = 0{\cdot}6 \text{ in}^3/\text{s}$$

(13) The flow for each pocket is calculated.

$$q' = \frac{q_o}{n} = \frac{0{\cdot}6}{6} = 0{\cdot}1 \text{ in}^3/\text{sec}$$

(This value is used for calculating the control device).
(14) The flow is converted to familiar British units.

$$q_o = 0{\cdot}6 \times \frac{60}{277} = 0{\cdot}13 \text{ gal/min}$$

(15) The value for the tilt parameter $\bar{K}$, for the design condition, is read from Fig 5 for the values $\bar{D} = 1{\cdot}58$, and $\bar{P} = \beta = 0{\cdot}5$. Thus, $\bar{K} = 0{\cdot}175$
(16) The tilting stiffness is calculated from equation 4.

$$k_o = \frac{P_s A D_o^2}{2 h_o} \bar{K}$$
$$= \frac{300 \times 8{\cdot}9 \times 4{\cdot}34^2}{2 \times 0{\cdot}002} \times 0{\cdot}175$$
$$= 2{\cdot}2 \times 10^6 \text{ lbf in/rad}$$

(17) The maximum angle of tilt before metal to metal contact at one edge is

$$\theta = \frac{2h_o}{D_o} = \frac{0{\cdot}004}{4{\cdot}34} = 0{\cdot}00092 \text{ radians}$$

(18) The design value of maximum tilting load is calculated.

$$M = \tfrac{1}{2} k\, \theta = \tfrac{1}{2} \times 2{\cdot}2 \times 10^6 \times 9{\cdot}2 \times 10^{-4}$$
$$= 101 \text{ lbf in.}$$

(19) The value of $\bar{\lambda}$ for $\bar{P} = 0{\cdot}5$ and capillary control with $\beta = 0{\cdot}5$ is read from Fig 4(*a*). $\bar{\lambda} = 0{\cdot}75$.
(20) The axial stiffness is calculated using equation 3.

$$\lambda = \frac{P_s A \bar{A}}{h_o} \bar{\lambda} = \frac{300 \times 8{\cdot}9 \times 0{\cdot}75}{0{\cdot}002} \times 0{\cdot}75$$
$$= 7{\cdot}52 \times 10^5 \text{ lbf/in}$$

(21) The maximum overload condition is calculated. $W_{\text{max}} = 1200$ lbf
(22) The value of $\bar{P}$ at this condition is calculated.

$$\bar{P} = \frac{W_{\text{max}}}{W_o} \times \beta = \frac{1200}{1000} \times 0{\cdot}5 = 0{\cdot}6$$

(23) The value of $\bar{K}$ at this condition is read from Fig 5. $\bar{K} = 0{\cdot}17$
(24) The value of $\bar{X}$ is read from Fig 6*a* for $\bar{P} = 0{\cdot}6$, and $\beta = 0{\cdot}5$. $\bar{X} = 0{\cdot}86$
(25) The value of h is calculated.

$$h = h_o \bar{X} = 0{\cdot}00172 \text{ in}$$

(26) The tilting stiffness is calculated.

$$k = \frac{P_s A\, D_o^2}{2h}\bar{K}$$
$$= \frac{300 \times 8{\cdot}9 \times 4{\cdot}34^2}{2 \times 0{\cdot}00172} \times 0{\cdot}17$$
$$= 2{\cdot}5 \times 10^6 \text{ lbf in/rad}$$

(27) The maximum angle of tilt is calculated.

$$\theta_{max} = \frac{2h}{D_o} = \frac{2 \times 0{\cdot}00172}{4{\cdot}34} = 0{\cdot}0008 \text{ radians}$$

(28) Maximum design tilting load at overload is

$$\tfrac{1}{2}k\,\theta_{max} = \tfrac{1}{2} \times 2{\cdot}5 \times 10^6 \times 8 \times 10^{-4}$$
$$= 100 \text{ lbf/in}$$

(29) The value of $\bar{\lambda}$ for $\bar{P} = 0{\cdot}6$ is read from Fig 4(a),

$$= 0{\cdot}96$$

(30) The axial stiffness at overload is calculated from equation 3,

$$\lambda = \frac{P_s A\, \bar{A}}{h_o}\bar{\lambda} = \frac{300 \times 8{\cdot}9 \times 0{\cdot}75}{0{\cdot}002} \times 0{\cdot}96$$
$$= 9{\cdot}6 \times 10^5 \text{ lbf/in}$$

Speed effects in annular thrust bearings

Unlike the journal bearing, the tilted annular thrust bearing has a poor geometry for induced hydrodynamic effects. Consequently it is not as important to observe the optimum conditions for minimum power, since the undesirable effects of hydrodynamic action will occur at a much higher speed than the optimum for this case. However, for minimum power loss it is necessary to keep the value of the ratio of H_f/H_p between 1 and 3 as before for a cool bearing. The pumping power is:

$$H_p = P_s q = \frac{P_s^2\, h_o^3}{\eta}\beta\,\bar{B}$$

The friction power loss is

$$H_f = \eta\, A_f \frac{V^2}{h_o} = \frac{\eta\, D_o^4\, N^2}{h_o} \times \pi^2 \left(\frac{A_f}{D_o^2}\right)\frac{(D_o + D_i)^2}{(2D_o)^2}$$

Equating these expressions, such that $H_f = H_p$,

Fig 5

Curves for determining the stiffness factor $\bar{K}$

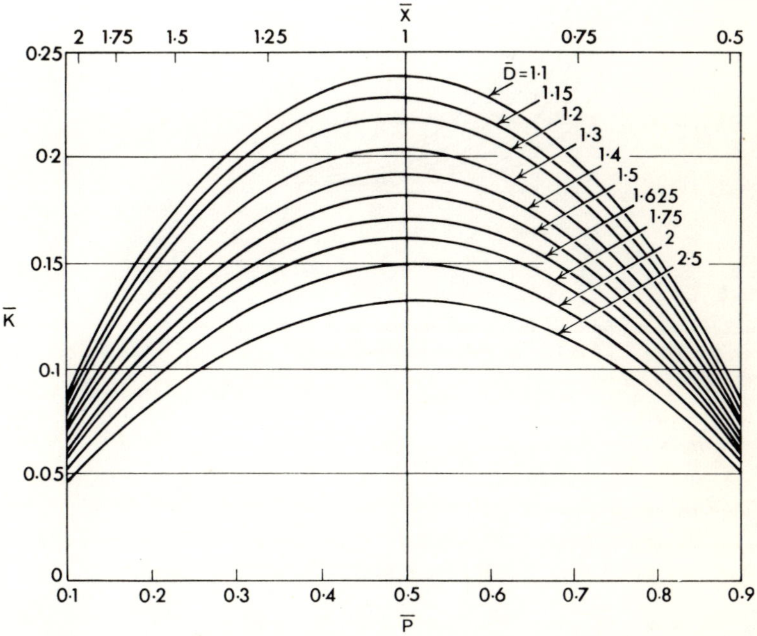

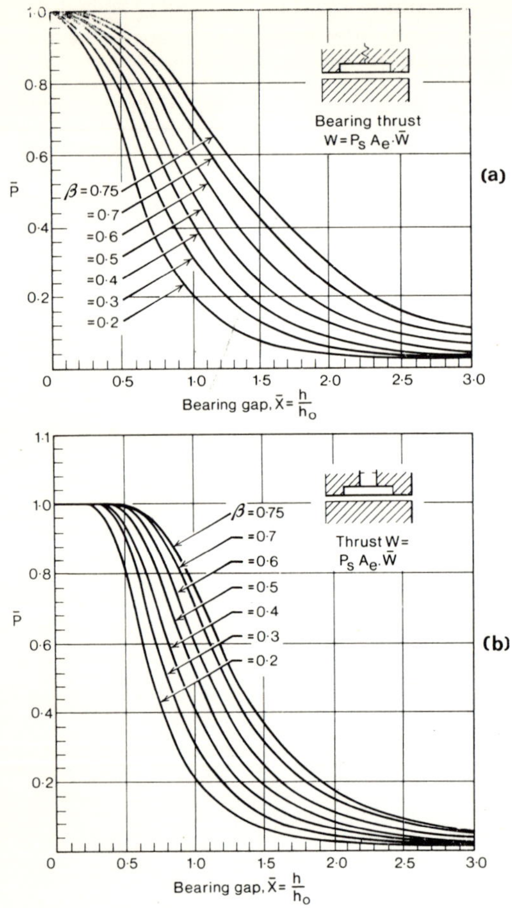

Fig 6

The bearing gap $\bar{X}$ for a capillary-controlled bearing is taken from the curves at (a) and for an orifice-controlled bearing from the curves at (b)

$$\frac{\eta\, D_o^4\, N^2}{h_o} \times \pi^2 \left(\frac{A_f}{D_o}\right) \frac{(D_o + D_i)^2}{(2D_o)^2} = \frac{P_s^2\, h_o^3}{\eta}\, \beta\, \bar{B}$$

and re-arranging,

$$\frac{\eta^2 N^2}{P_s^2}\left(\frac{D_o}{h_o}\right)^4 = \frac{\beta\,\bar{B}}{\pi^2 \frac{A_f}{D_o}\left(\frac{D_o + D_i}{2D_o}\right)^2}$$

For $\bar{c} = 0\cdot25$, $\beta = 0\cdot5$

$$S_H = \frac{\eta\, N}{P_s}\left(\frac{D_o}{h_o}\right)^2 = \frac{1}{\pi}\sqrt{\frac{\bar{B}}{1\cdot25\left(1 - \frac{1}{\bar{D}^4}\right)}}$$

Now if $H_f = 3H_p$ $\quad S_H = \frac{\sqrt{3}}{\pi}\sqrt{\frac{\bar{B}}{1\cdot25\left(1 - \frac{1}{\bar{D}^4}\right)}}$

For the above example, suppose the speed $N = 500$ rev/min, then the optimum value of viscosity, η, could be found from the relationship,

$$S_H = \frac{1}{\pi}\sqrt{\frac{\bar{B}}{1\cdot25\left(1 - \frac{1}{\bar{D}^4}\right)}}$$

$$= \frac{1}{\pi}\sqrt{\frac{12\cdot5}{1\cdot25\left(1 - \frac{1}{1\cdot58^4}\right)}}$$

$$= 1\cdot1$$

So that

$$\eta = S_H \times \frac{P_s}{N}\frac{h_o^2}{D_o^2}$$

$$= 1\cdot1 \times \frac{300 \times 60}{500} \times \left(\frac{0\cdot002}{4\cdot34}\right)^2$$

$$= 1\cdot1 \times 36 \times 4\cdot6^2 \times 10^{-8}$$

$$= 8\cdot4 \times 10^{-6} \text{ reyn}$$

Temperature rise

$$\Delta T = 0\cdot015\, P_s$$

$$= 4\cdot5°\text{F}$$

In this case, the angle of tilt is displaced 8° from the angle of applied tilting moment, with a slight increase in tilting resistance. In optimized cases, the effect of hydrodynamic action can safely be neglected.

REFERENCES

1 J P O'Donoghue and W B Rowe 'Hydrostatic bearing design', *Tribology Vol 2 No 1, February 1969, pp 25-71*

Chapter Nine

A basis for inter-active bearing design procedures

J. P. O'DONOGHUE and W. B. ROWE

A number of simplified design procedures has been presented for the most usual hydrostatic bearing configurations. The basic procedures are self-contained and cross-referencing has been largely avoided, although this practice has led to a certain amount of duplication. However, the prime objective has been that the designer should not have to spend an excessive period in assimilating ideas from a mass of information. Very often it will be sufficient for the designer to adopt the standard approach without much analysis of fundamental principles.

The purposes of this chapter are:

1. To emphasize the common underlying basis of the procedures.
2. To indicate how the procedures may be modified by a designer if his requirements do not correspond exactly with the standard approach.
3. To indicate where problems may be encountered as a result of departing from the suggestions contained in the design procedures. The procedures tend to be conservative and sometimes over-safe, and if it is decided to design less 'conservatively' it is advisable that the designer should read this chapter carefully.

Although the procedures form a rational approach to design, the same basic relationships may be rearranged and recombined in a versatile manner where required. For example, in the simplified journal bearing procedure, optimization is achieved by calculation of viscosity, and it may be necessary to alter previous decisions on geometry or clearance to arrive at a suitable value. However, in some machinery there may be no choice concerning viscosity value and in such a case it is not difficult to rearrange the optimization equation to yield a value of optimum clearance.

The general basis of the design procedures is summarized in Fig 1, which should illustrate the point being made.

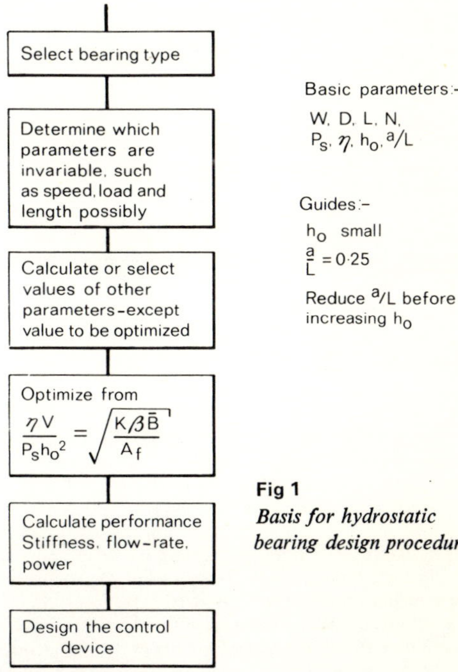

Fig 1
Basis for hydrostatic bearing design procedures

Compensation method

A hydrostatic bearing maintains a stable oil film by virtue of the associated control device. Essentially, this device serves to alter the flow-rate to compensate for variations of bearing load, and the combined effect of the latter factor and the flow-rate determine what the deflected bearing position shall be. Control devices and their operation have been described fully[1]. The principal types are below discussed in order of increasing bearing stiffness. (They are summarized in Fig 2.)

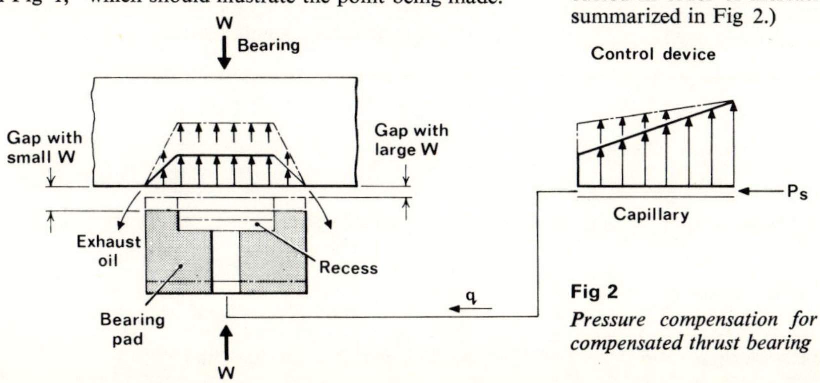

Fig 2
Pressure compensation for applied load in a capillary-compensated thrust bearing

VISCOUS RESTRICTORS (e.g. capillaries, slots, annuli, screw-threads). An advantage of a viscous restrictor is that the gap is independent of temperature for as long as viscosity in the restrictor is equal to viscosity in the bearing.

Flow-rate through a viscous resistance R follows a relationship of the form

$$q = \frac{(P_s - P_r)}{R}$$

and for a capillary the resistance

$$R = \frac{128\,\eta l}{\pi d^4}$$

Viscous restrictors tend to be less prone to blockage than orifice (turbulent) restrictors, because their critical dimensions are larger. In designing viscous restrictors, care should be taken to ensure (a) a sufficient length of flow-path to avoid entrance and exit effects (preferably $l > 100 \times$ smallest dimension diameter or clearance); and (b) Reynolds number < 1000.

A design procedure for capillary restrictors is given in chapter 1, and may be employed for bearings of any type. The procedure is appropriate for any viscous restrictor if an appropriate equation is substituted in instruction (4), which as it stands reads:

$$l = \frac{\pi d^4 \; K_c}{128}.$$

K_c is the capillary resistance constant.

TURBULENT RESTRICTORS (e.g. orifices) The principal advantages of the orifice-type restrictor as compared with the capillary form are slightly improved stiffness, and economy of space (since l is of the same order as d, and preferably $l/d << 1$). It should not be assumed that increased stiffness results in increased load-bearing capacity. In fact, high stiffness in the design conditions may slightly reduce load-bearing capacity.

The disadvantages of the orifice restrictor are increased susceptibility to blockage, and a temperature dependence which is of importance in machine tools and measuring equipment. If the viscosity changes owing to a temperature variation, there will be a change in bearing position. However, an effect which tends to occur in bearings operating over a wide temperature range is 'thermal collapse'. Thermal collapse problems are less severe with orifice-controlled bearings than with viscous restrictors, and the same comment applies to constant flow devices.

Flow-rate through a turbulent restrictor follows a relationship of the form

$$q = \frac{1}{s}\sqrt{(P_s - P_r)}$$

where s is dependent on Reynolds number.

A design procedure for orifice restrictors is also given in chapter 1, and may be employed for bearings of any type.

CONSTANT FLOW PUMPS AND CONSTANT FLOW VALVES. One pump or valve is required for each recess, and the principal advantages of pump control are increased stiffness, reduced heat, and a negligible tendency to blockage. In the case of valve control, only the first-mentioned advantage applies, namely increased stiffness. If there is only one pad in the system then a constant flow pump or valve would appear to be the ideal arrangement.

Design is very simple since the required flow-rate is known after carrying out the pad design calculations.

PRESSURE-SENSING AND POSITION-SENSING VALVES. Several valves of these types have been described. Suitable design ensures practically infinite stiffness over a large proportion of the load range, and in certain applications (such as high precision machine tools) these devices may be justified in spite of their increased complexity and cost.

Basic relationships

LOAD. Load may be calculated for any bearing from the following equation:

$$W = P_s A.\bar{A}.\bar{W}$$

or alternatively from

$$W = P_s A_e \bar{W}$$

For a single plane pad, this simplifies to

$$W = P_r A_e = P_s A_e \bar{P}$$

The effective area of a bearing depends only on the bearing type and shape. Values of $\bar{A}$ or A_e may be read from charts or tables of the type included in each of the design procedures described in previous chapters or elsewhere[2].

Symbols

P_s	Supply pressure	$\bar{X}$	Film thickness ratio (h/h_o)
P_r	Recess pressure	λ	Bearing stiffness
$\bar{P}$	P_r/P_s	$\bar{\lambda}$	Dimensionless stiffness ($\lambda h_o/P_s A_e$)
h	Film thickness	q	Flow rate
$c = h_o$	Normal clearance	$\bar{Q}$	Dimensionless flow rate ($q\eta/P_s h_o^3$)
C_D	Diametral clearance ($2h_o$)	H	Power
D	Bearing diameter	K	Ratio of friction power to pumping power
L	Bearing length	ρ	Mass density of oil
a	Flow land width from recess to exhaust	S_H	Optimization parameter
β	Design pressure ratio (P_o/P_s)	C	Specific heat of oil
η	Viscosity	ΔT	Maximum temperature rise between oil inlet and oil outlet
A	Projected bearing area	d	Capillary diameter
A_e	Effective bearing area	l	Capillary length
A_f	Effective friction area	K_c	Capillary constant ($128l/\pi d^4$)
V	Sliding speed	S	Orifice factor
N	Rev/s	J	Mechanical equivalent of heat
$\bar{W}$	Dimensionless load $= W/P_s.A_e$	R_e	Reynolds number
$\bar{B}$	Computed flow factor for pad	o	Suffix denoting design condition

The load parameter $\overline{W}$ depends on the recess pressures, and hence on bearing gap and the type of control device. Values of $\overline{W}$ (or $\overline{P}$ as appropriate) are read from charts for the given bearing application. The suggested values of $\overline{W}$ are conservative in most cases, so that with a little care it is sometimes possible to increase $\overline{W}$ slightly. For example, with journal bearings it is suggested in effect that $\overline{W}_{max} = \bar{\lambda}_o^*/3$. With capillary control, however, and $\beta = 0{\cdot}5$ under optimum conditions it is safe to operate when $\overline{W}_{max} = \bar{\lambda}_o^*/2$.

STIFFNESS. Stiffness is the rate of variation of bearing load with deflection, and tends to vary with load. As the ultimate load of a bearing is approached, there is a rapid decline in stiffness, which is one reason why a margin of safety should be allowed in determining the maximum load.

Stiffness for any bearing may be calculated from the following equation:

$$\lambda = \frac{P_s A^* \bar{A}^* \lambda}{h_o}$$

or alternatively,

$$\lambda = \frac{P_s A_e \bar{\lambda}}{h_o}$$

The values $\bar{A}$ and A_e are identical to those used previously for load. Values for $\bar{\lambda}$ depend on the control device, and are provided on charts for specific bearing applications.

FLOW RATE. Flow rate tends to vary considerably with temperature as a result of viscosity and clearance variations. Therefore accurate predictions are only possible if the viscosity and clearance are accurately determined. The flow rate may then be determined from:

$$q = \frac{P_s h_o^3}{\eta} . \bar{Q}$$

where $\bar{Q} = \bar{P}. \bar{B}. \bar{X}^3$.

For single plane pads, this relationship is simplified to:

$$\bar{Q}_o = \beta\bar{B},$$

in the design condition when $h = h_o$ and by definition, $\bar{P}_o = \beta$.

The flow rate through a journal bearing is usually at a maximum with the concentric condition, when $h = h_o = c$, so that:

$$q_o = \frac{P_s c^3}{\eta} . \bar{Q} .$$

Values for $\bar{Q}$ may be read from charts or calculated from a given expression, depending on the particular bearing application.

POWER DISSIPATION. Optimization for low power dissipation is particularly important, and excessive dissipation results in increased cost for the ancillary equipment and leads to thermal problems. Moreover unless a bearing is optimized, there is a tendency for the behaviour to fall short of the ideal inherent in hydrostatic bearings. Some of these effects will be apparent from the later sections of this chapter, dealing with optimization, thermal effects, hydrodynamic effects and tolerances.

Heat is generated due to pumping power dissipation (H_p) and to sliding friction in the oil film as a result of relative bearing motion (H_f). The total power is the sum of H_p and H_f. Thus,

$$H_t = P_s q + \frac{\eta A_f V^2}{h}$$

or $H_t = (K + 1) P_s q$.

This relationship applies for an optimized bearing as explained later, in the section dealing with optimization.

EFFECTIVE FRICTION AREA. The effective friction area is defined as that area over which sliding with constant film thickness h gives rise to the same friction as the actual bearing, including the lands and recesses. Friction in the recesses is reduced by the greater recess depth, but not as much as would be expected because of churning and recirculating effects. It has been found that if the recess depth is greater than the bearing clearance by a factor of 20, an estimated value for bearing friction is obtained by writing:

$$A_f = A - \tfrac{3}{4} A_{recess}$$

In some of the following procedures, an optimization parameter S_H has been introduced which avoids complexity in the calculations. In these cases, it is not necessary to calculate the friction area, since this factor is taken into account by the computed values for S_H.

Optimization of bearings operating at speed

The minimum power for any required operating condition may be determined by differentiating the expression for

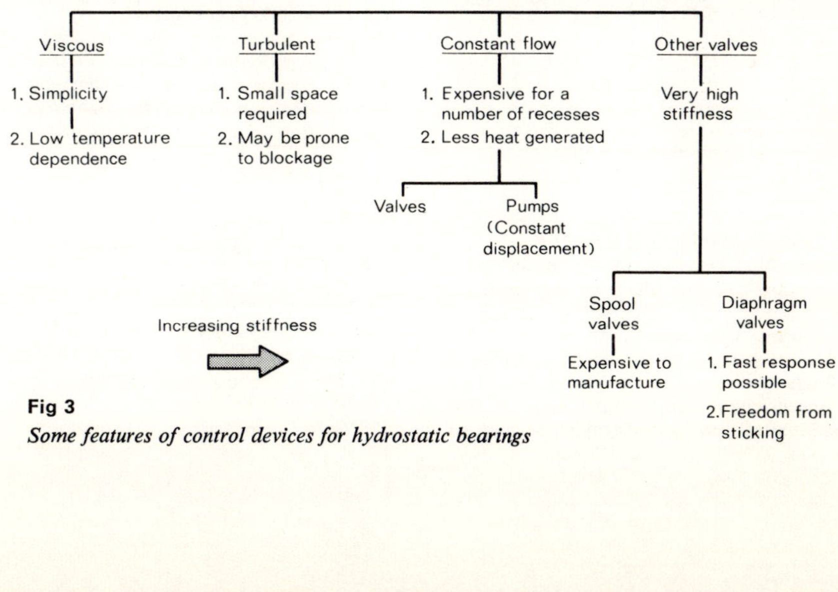

Fig 3

Some features of control devices for hydrostatic bearings

total power and equating to zero. For example, given a range of operating conditions such as speed, size, and clearance, it is possible to determine the optimum viscosity. It is found that irrespective of which parameter is optimized, operation will always be within 15 per cent of optimum for the required operating conditions if:

$H_f = KH_p$,

where $1 < K < 3$,

so that $H_t = (K+1)\, H_p$

If a tolerance has to be held, then the range of K may be useful to set the limits. The procedures are usually directed towards the condition $K = 1$, because of thermal and hydrodynamic considerations. If bearings are optimized for the free variable after constraining all other values, the following conclusions are reached.

1. The lowest power condition of all is obtained with minimum clearance, a value of approximately 0·25 for the land-width ratio (land width/bearing width), and optimization performed with respect to viscosity.
2. If this procedure is repeated with a smaller land-width ratio, power will be slightly increased and viscosity will be increased. The arrangement may consequently be of advantage where a higher viscosity must be employed.
3. Increasing clearance also has the effect of increasing optimized viscosity and power. If the objective is higher optimized viscosity, it is preferable to reduce the land-width ratio before increasing the clearance.

If after determining the optimum viscosity, the value is found to be considerably higher than is to be used in practice, the result may be interpreted as an indication of a slow speed bearing. In this case, optimization is unimportant, and no larger error will result by writing $K = 0$.

Below are shown the results of some calculations concerning a typical plane pad bearing, and they are compared graphically in Fig 4. Initially, values of all variables were selected for convenience, and those chosen for viscosity and clearance were the minimum that are considered convenient. Solutions 2 to 6 all involve operation at relatively low power and temperature. If the viscosity must be maintained at the initial value, solutions 5 and 6 indicate a suitable working range.

	η micro-reyns	a/L	h_o in	K	H_t lbf in/sec	ΔT (relative)
1	8	0·25	0·0005	40	39·8	20
2	1·3	0·25	0·0005	1	12·6	1
3	2·26	0·25	0·0005	3	14·7	2
4	8	0·25	0·00094	3	27·4	2
5	8	0·06	0·0005	3	19·8	2
6	8	0·06	0·00066	1	22·6	1

The optimization condition may be expressed in terms of the variables, and results in the following equation:

$$\frac{\eta V}{P_s h_o^2} = \sqrt{\frac{K\beta\bar{B}}{A_f}}$$

This equation applies generally and usually $K = 1$, although as previously discussed, the value may be increased to 3 with only small effect. To determine the optimum value of any one variable, it is merely necessary to substitute the other values into this equation.

As a guide to the other variables it must be remembered that clearance should be as small as possible for high stiffness and low power dissipation. Achievement of this tendency is limited by tolerances which must be applied

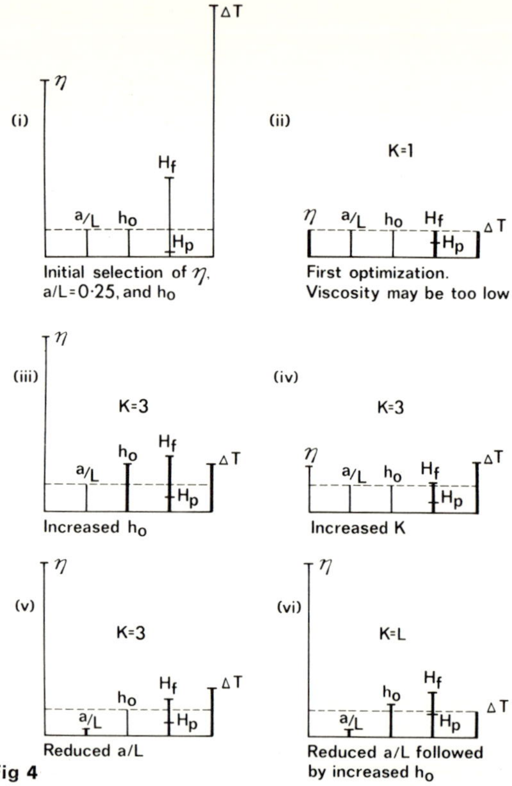

Fig 4

Charts concerning alternative designs for a bearing showing the effects on power and temperature rise (thick lines indicate the results of the optimization process)

as regards size and accuracy, and the values should be small in comparison to clearance – namely, less than 20 per cent. Speed also forms a limiting factor, when it is found that even after reducing land widths, the viscosity indicated is too low for an application. The land-width ratio should be 0·25, unless it is necessary to reduce the value to increase the viscosity.

The optimization parameter

The optimization parameter S_H was not employed in all the design procedures, but with familiarity, a designer will find the parameter an aid to establishing optimum designs. The parameter is derived by slight rearrangement of the optimization relationship. For plane bearings:

$$\frac{\eta L V}{P_s h_o^2} = S_H,$$

where

$$S_H = \sqrt{\frac{K\beta\bar{B}}{A_f}}$$

Computed values for S_H, relating to a value of 0·5 for β, are indicated graphically at (a) in Fig 5. The advantage for a designer is that it is simple to determine which value for land-width ratio to use for a precise viscosity value, and a considerable reduction in arithmetic results in practice. For journal bearings:

Fig 5
Values for the optimization parameter S_H, an optional aid to optimizing bearings which operate at speed ($\beta = 0{\cdot}5$), are obtained for plane bearings from diagram (a), and for journal bearings, from that at (b)

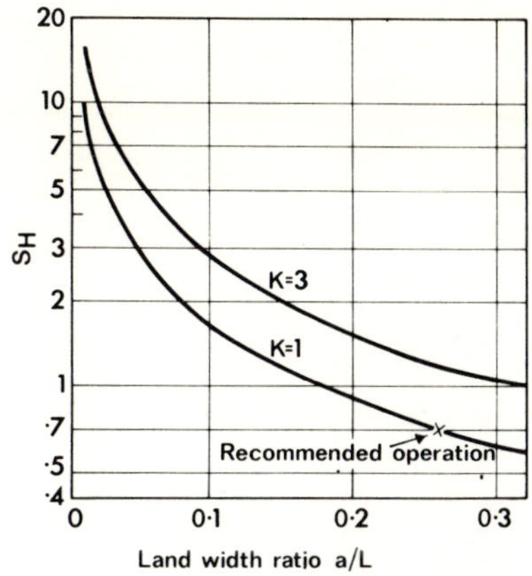

(a)

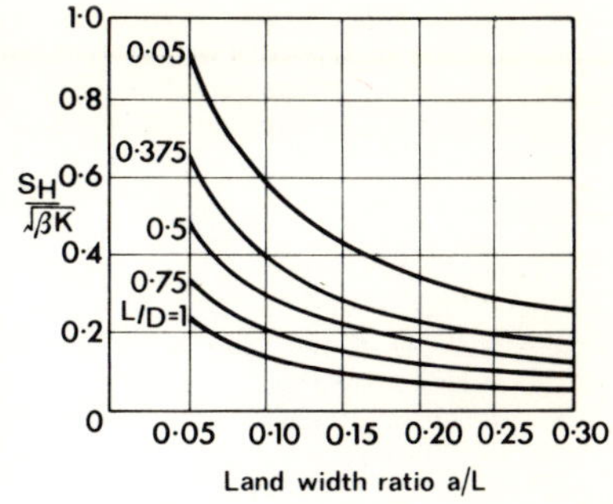

(b)

$$\frac{\eta N}{P_s}\left(\frac{D}{C_D}\right)^2 = S_H$$

A general chart with computed values is given at (b) in Fig 5.

Design pressure ratio

In most cases there is little to be gained by using a value other than 0·5 for β. Slightly increased stiffness is sometimes obtained in the concentric position, but from experience of a number of considerations, the following are the principal applications for which there are advantages in alternative β values.

1. Single plane pads. Where the extreme load on a pad is much higher than the normal operating load, an improved load range may be obtained by increasing the bearing area and reducing the design pressure ratio. The reverse procedure may be applied where the load is to be greatly reduced from the design value.
2. Primary and secondary hydrostatic bearings. Where the efflux from a primary hydrostatic bearing forms the pressure supply to a secondary hydrostatic bearing, the usual procedure will be to increase the pressure ratio of the primary bearing and reduce the pressure ratio of the secondary bearing.
3. Complementary pairs of conical journal bearings. By reducing the pressure ratio from 0·5 to 0·3, it is possible to obtain an improved range of combined radial and axial loading.

Thermal effects

The complete prediction of thermal effects is complex, as may be apparent from a consideration of the ways in which heat is generated and dissipated in oil as it passes through a capillary-controlled bearing

With such a bearing, oil enters the capillary at a temperature which depends on the characteristics of the supply system, and the value may be well above the ambient temperature. Passing through the capillary, the oil is losing pressure and therefore generating heat. Heat is also conducted away from the oil through the capillary walls, dependent on the temperature difference above ambient. However, oil is a relatively poor conductor of heat and therefore most of the generated heat is normally retained in the oil. Due to the resultant temperature rise, the viscosity is reduced. Subsequently, the oil enters the bearing recess, on the way to the bearing land, and at this stage there is negligible drop in pressure due to the recess depth, although some heat passes by conduction to both bearing members. On the other hand, if the bearing is operating at speed, there will be friction losses. In a third stage, when the oil passes through the clearance between the bearing lands, further heat is generated due to pumping and friction losses, and a proportion of this heat is conducted away through the bearing members.

At low speeds, low pressures, and low flow-rates, the overall result of this series of conditions may be cooling of the oil as it passes through the bearing. However at high speeds, high pressures and high flow-rates, it is much more likely that the temperature will be rising and the viscosity progressively falling. A further complication is that expansion of bearing members may reduce the clearance and increase the tendency to heating. When considering the severity of the thermal problem, a designer should decide whether the application lies in one of the following areas.

1. Stationary bearing, moderate pressure (< 1000 lbf/in^2).
2. Optimized sliding bearing, moderate pressure.
3. Stationary bearing, high pressure.
4. Sliding bearing, high pressure and/or super-critical speeds.

The design procedures provide for the first two cases, but the third and fourth cases require a more complex analysis, as will be apparent from the following discussion.

STATIONARY BEARING, MODERATE PRESSURE. Heat is generated due to dissipation of the pumping power. The maximum theoretical temperature rise may be calculated by assuming constant viscosity. In practice there may be a temperature drop due to heat conduction. The power dissipation is related to the maximum temperature rise, according to:

$$H_p = P_s q = J \rho q C \,.\, \triangle T.$$

Hence, $$\triangle T = \frac{P_s}{J \rho C}$$

For light machine oils, it is found that the maximum temperature rise is $\frac{3}{4}$°F per 100 lbf/in^2 supply pressure.

Optimized sliding bearing, moderate pressure. From the optimization process, it is established that:

$$H_t = (K+1) H_p = (K+1) P_s q$$

Hence, the maximum theoretical temperature rise for constant viscosity is given by:

$$\Delta T = \frac{(K+1) P_s}{J \rho C}$$

For light machine oils with $K = 1$, the result of this expression corresponds approximately to a maximum temperature rise of 1½°F per 100 lbf/in² supply pressure, and with $K = 3$, the value is increased to 3°F per 100 lbf/in² supply pressure.

At moderate pressures (the design procedures recommend 300 lbf/in²) the effect of viscosity variations and heats conduction tend to offset each other, so that serious problems do not arise with conservative procedures. In this connection, it is worth mentioning that very thin oils do not vary in viscosity so much as thick types. The optimization process tends towards very thin oils for high-speed applications, so account is taken automatically of this feature.

Stationary bearing, high pressure. With high working pressures, the viscosity is increased at a bearing inlet, due to the pressure, but it is reduced at the outlet due to temperature increase which occurs. The variation of viscosity with pressure and temperature may be expressed approximately by:

$$\eta = \eta_0 e \frac{(\alpha P - \beta T)}{\text{exponent}}$$

The balance of heat at any point may be expressed in the following form:

$$\frac{1}{J} (q.dp) = \rho q C \Delta T + \frac{K.\, dA(T_a - T_o)}{L_H}$$

The temperature effect is best offset at low flow-rates, for with such conditions, a relatively large proportion of the heat is conducted away. At high flow-rates due to excessive clearance, there will be a progressive reduction in viscosity. The effect of the reduction in viscosity is to reduce the minimum film thickness. It may be noted that the second term on the right hand side of the heat balance equation may sometimes be neglected without serious loss of accuracy since with high flow-rates, the value is small by comparison with the other terms.

Sliding bearing, high pressures and/or super-critical speeds. The heat balance equation must include the friction term, so that in this case, the expression takes the following form:

$$\frac{1}{J}\left(\frac{\eta V^2.dA}{h} - q\,.\,dp\right) = \rho q C \Delta T + \frac{K.dA(T_a - T_o)}{L_H}$$

If all speeds up to that for which a bearing was optimized are referred to as sub-critical, then higher speeds are super-critical. Under such conditions, an analysis based on varying viscosity with pressure and temperature, together with application of the above heat balance equation, indicates the possibility of thermal collapse. The basic manner of this collapse is that due to the progressive reduction in viscosity through a bearing the pressure distribution is so altered that a higher recess pressure is required to carry a given load. As the recess pressure approaches the maximum, the value cannot be increased sufficiently to support the load, and collapse occurs. Even before this stage is reached, however, there is reduced film thickness as a result of the viscosity reduction, which may lead to failure. Curves indicating these points are shown in Fig 6, and they are based on computed results, using typical values for conductivity for a rotating circular thrust pad. The collapse is not so serious with thin-land bearings as with those having thick lands, but the reduction in film thickness effect still occurs. The critical speed for the example given in Fig 6 is 800 rev/min.

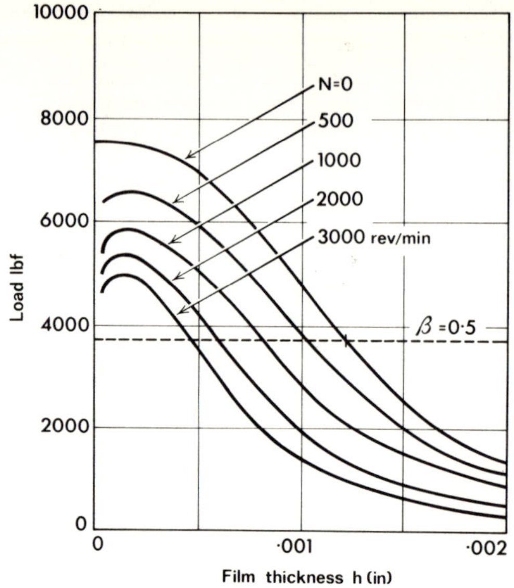

Fig 6

Typical computed load and stiffness characteristics against are here plotted in relation to clearance, taking thermal effects into account. The data concern a rotating thrust bearing, 3 in recess diameter, 6 in outside diameter, for which the critical speed is 800 rev/min

Hydrodynamic effects

When one bearing member slides relative to a fixed member, fluid is transferred in the direction of movement. If the two members are parallel, this action has no effect, except in the following cases.

1. The leading edge of the moving member may be starved at super-critical speeds (where the critical speed is the speed at which the bearing was optimized).
2. If the clearance at the leading edge is larger than at the trailing edge, hydrodynamic lifting will occur in addition to the normal hydrostatic lifting, due to the 'wedge' effect. However, if the member is tilted in the opposite sense, so that clearance at the trailing edge is larger, the opposite effect occurs and there is a tendency to collapse. An analysis of a hydrostatic bearing which has been optimized ($K = 1$) but is tilted by 20 per cent indicates that the hydrodynamic lifting effect is of the order of 4 per cent of that obtained hydrostatically.

Hydrodynamic effects also occur in journal bearings. A hydrodynamic thrust is exerted at an angle of 90° to the line of eccentricity. As with other effects, the performance of a bearing is not seriously affected at sub-critical speeds. In an optimized journal bearing ($L/D = 1$) it is found that the line of eccentricity is inclined at an angle of approximately 35° with respect to the applied load line. This effect is approximately constant up to an eccentricity ratio of 0·6, and the increase in lift obtained in this way is less than 20 per cent.

A further consequence of the hydrodynamic effect is cavitation at high values for eccentricity. Cavitation involves the absence of fluid, and results from the low pressures on the lightly-loaded negative wedge region of a bearing. It serves to reduce the dynamic stiffness of a bearing, and is therefore undesirable. At super-critical speeds, the onset of cavitation occurs at relatively lower eccentricity values. At the values which would otherwise be the maximum normally acceptable for eccentricity, a super-critical bearing rapidly tends to become an inefficient hydro-dynamic type with increasing speed.

Turbulence

The design procedures are based on laminar flow of the fluid in a bearing. Turbulent conditions occur under conditions of large clearance, high viscosity, and high flow-rates, whether these factors are pressure- or velocity-induced. The conditions for turbulence will exist in a bearing recess at much lower speeds than in the region of the lands because the clearance is much larger.

A guide to the speed at which the transition occurs from laminar to turbulent flow, in journal bearings, may be obtained from the Taylor equation which is:

$$\pi D N_t \frac{\rho}{\eta} = 41{\cdot}1\sqrt{\frac{D}{2h}}$$

where N_t = journal transition speed, rev/sec.

Tolerances

Limits of size and shape must be determined, which will in turn affect the manufacturing processes to be employed. The principal performance considerations are:

1. Required accuracy of movement (related to shape errors).
2. Permissible variation in flow-rate.
3. Possible effects on load-carrying capacity, minimum film thickness, and stiffness.

There are many possible combinations of errors, so that it is not possible to give guidance on the allowable contribution from each error. Some analysis of the effect of various errors is given in reference 3. Probably the simplest way of affording guidance to a designer is to state that the summation of all errors must be within a given tolerance band. This statement may be interpreted as meaning that all points on the two bearing surfaces must lie within the region dictated by the tolerance. It is then for the designer to decide how he wishes to specify limits and whether the tolerances are to be shared between such factors as size, roundness, and straightness at the design stage, or whether the overall result will be assessed after inspection. Fig 7 shows the effect on operating eccentricity of variations in clearance in a journal bearing for an assumed value for the pressure ratio. The curves indicate that for a pressure ratio not exceeding 0·5 it is preferable that clearance be undersize rather than oversize. This conclusion agrees with optimization consideration, since if $K = 1$ for the largest value for clearance, then other values for clearance will tend towards $K = 3$, which is still in the optimum range.

Bearing in mind all the previous considerations, it is recommended that the total tolerance should not exceed 20 per cent of the maximum clearance.

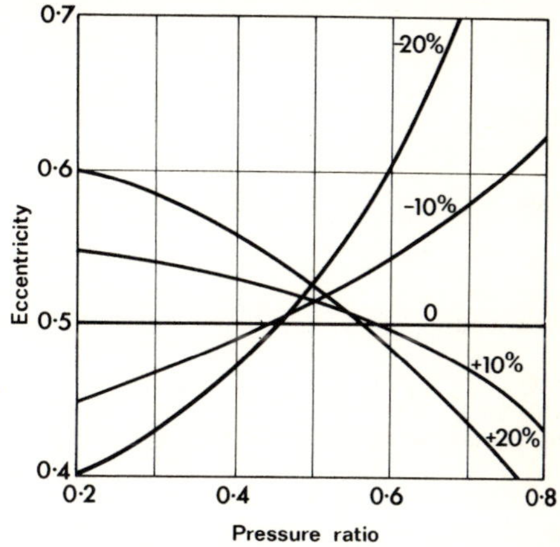

Fig 7

Variation of journal eccentricity in accordance with a given error in clearance an assumed value for the pressure ratio

REFERENCES

1. J. P. O'Donoghue and W. B. Rowe 'Compensation Methods for Externally Pressurised Bearings', *Engineers Digest*, Vol. 30, No. 4, April 1969, pp. 49-61.
2. J. P. O'Donoghue and W. B. Rowe 'Hydrostatic Bearing Design' *Tribology*. Vol. 2, No. 1, February 1969, pp. 25-71.
3. J. P. O'Donoghue, W. B. Rowe and C. J. Hooke 'Some Tolerancing Effects in Hydrostatic Bearings' *Presented at the 11th International Machine Tool Design and Research Conference*, September, 1970. Proceedings published by Pergamon Press.

Index

All references are to chapter numbers